MW00769072

"SO THERE WE WERE..."

**Leadership Stories from the Men and Women
Who Make the Navy Work**

*So: Heidi —
Love, Wendi
(signature)*

"SO THERE WE WERE..."

Leadership Stories from the Men and Women Who Make the Navy Work

Written by

Patricia Reily, Ed.D.

EMERGENT™

PUBLICATIONS

3810 N 188th Ave
Litchfield Park, AZ 85340

The photo on the cover is the work of John Hewick—a true renaissance man.

The photo of the anchor insignia is the work of Jim Reily—the love of my life.

The Broadside cartoons are the work of Jeff Bacon—a master of Navy humor.

Half of the royalties from each book go to the CPO Scholarship Fund established for the family members of Navy Chiefs by Robert J. Walker, Master Chief Petty Officer of the Navy, retired.

"So There We Were…": Leadership Stories from the Men and Women Who Make the Navy Work
Written by: Patricia Reily, Ed.D.

Library of Congress Control Number: 2011928221

ISBN: 978-0-9842165-7-4

Copyright © 2011 3810 N 188th Ave, Litchfield Park, AZ 85340, USA

All rights reserved. No part of this publication may be reproduced, stored on a retrieval system, or transmitted, in any form or by any means, electronic, mechanical, photocopying, microfilming, recording or otherwise, without written permission from the publisher.

Printed in the United States of America

This book is dedicated to the deck plate leaders who make the Navy work—The Chiefs.

And a big thanks to Jeff...

CONTENTS

PART I
THE NEXUS OF LEADERSHIP, ORGANIZATION AND STORY

PART II
USING STORY GENRE TO BUILD A STORY REPERTOIRE

PART III
LEADERSHIP NARRATIVES

PART IV
ORGANIZATIONAL ADAPTABILITY AND CREATIVITY THROUGH STORY

PREFACE

WHAT THIS BOOK IS ABOUT

This book is about the leadership principles and practices of a particularly adept group of leaders—Navy Chief Petty Officers (Chiefs)—the men and women who make the Navy work. But this book is about much more than leadership, it is about the essence of what makes organizations live or die. The premise is that there is a missed opportunity with respect to the use of story in organizations and that leaders who fail to recognize the importance of the stories that are circulating in their organizations do so at their peril. By examining the stories of Navy Chiefs a clearer understanding of the nature of leadership, stories and the more expansive narratives that surround stories will emerge. The thesis is that stories can be used to enhance the adaptability of any organization in a fast paced, ever-changing world. But even more importantly, by examining the stories of Navy Chiefs the reader will see how story and narrative are the very life-blood of organizations.

Stories about Navy leaders—mostly high ranking Commissioned Officers or war heroes—are legend both inside, and outside, of the Navy. And there are numerous biographies and studies of battlefield commanders that present compelling leadership models. But little has been written about the leadership practices of Navy Chiefs and no one has ever systematically studied how they work, yet they are responsible for accomplishing the bulk of the often unglamorous work that gets done in the Navy, usually under less than ideal conditions.

This book, developed from a systematic study of Navy Chiefs, provides an understanding of how they manage and lead. I argue repeatedly that every leader should be aware of the stories that have made them who they are and their organizations what they are today. Through authentic examples and exercises this book suggests ways to tap into the rich source of energy that stories can be as they travel through organizations connecting individuals in a web like fashion, energizing and giving life to an organization. The conclusion is that through increased awareness and understanding of narrative and how it works, leaders in the Navy—as well as leaders in other human organizations—can mindfully engage story, and the more expansive universe of narrative that surrounds stories to make sense of the past, cope in the present and successfully navigate into the future.

What makes this book different from other books about leadership and organization?

This book is more than a rehash of tried and true leadership principles. It presents an opportunity for a philosophical paradigm shift with respect to the way leaders and managers view the world and how organizations function. As humans we constantly strive to make sense of what is happening around us, often using metaphors to organize our thoughts and paradigms to frame our interpretations. The facts do not change but the way we organize them—the lens through which we choose to observe them—can change and lead to new insights and growth. Biological metaphors—in contrast with the machine metaphors that are often used—lend new insights into how organizations and the people who live in them prosper or decline. Stories, narratives, and human organizations in general, function more like living, breathing biological systems than machines. When we think about human organizations as biological systems we can begin to see the power and potential of story and narrative.

The way to understand organizations, as well as enhance and foster their creativity and adaptability to rapidly changing environments, is to understand the stories and the overarching narratives at play in organizations. Through understanding and interpreting organizational story, the people who comprise organizations can find better ways to adapt, thrive and fix problems.

Who should read this book?

This book will appeal to those interested in leadership and organizational behavior. The stories of Navy Chiefs will have particular relevance for those familiar with the Navy, but they will also resonate with leaders beyond the military because although the story details and characters are different in other organizations the plots and themes that are uncovered in organizational stories are surprisingly universal.

More importantly, this book will appeal to those who have found old explanations for how human organizations work lacking and are searching for new explanations for how organizations really work. The theories of Complexity Science—a non-linear, emerging, self-organizing way of viewing how living systems work—are the grounding for the interpretation of the stories included here. I hesitate even using the term Complexity Science because readers unfamiliar with that term immediately assume "complicated" and "boring," rather than "practical" and "inspirational," for a way of looking at the world that simply makes sense. But after trying other labels, I decided that I needed to stick with the terms that those far smarter than I have used to describe this non-linear way of interpreting how human systems work. If for no other reason, using the labels most widely used in academia may point those who are interested in learning more about Complexity Science in the right direction. (For a more detailed explanation of Complexity Science theory see Appendix: A)

A note on language

Sailors have long had a reputation for using "salty" language. That being said, the language of 18 to 25 year old sailors in the Navy is probably not much different than the language of their cohorts on college campuses across the U.S. And the informal discourse I have encountered in corporate America in recent years is just as "salty" as the language used informally in the Navy. Indeed, the language of sailors may be less offensive than that of their civilian counterparts because there has been a concerted effort in recent years to eliminate the use of obscenities from the Navy vernacular. The stories and dialogues included here do include some crude language. I apologize to those who are offended by the inclusion of crude language or references to crude behavior in a "professional book," but I hope that the reader will understand that to maintain authenticity the language of the speakers was retained to the extent that it could be. There were some obscenities that I did not feel comfortable repeating in a professional book, but for the most part the language used is reproduced here as it was spoken. Pseudonyms were used for the characters to maintain their privacy, but the stories and dialogues are to the greatest extent possible authentic.

This book is organized into four Parts. Part I sets the stage by: 1) briefly presenting the theories that inform the rest of the discussion in the book, and 2) describing the settings and characters—the context—for the stories, narratives and discussion that follows.

Part II uses a classification system and examples of stories to begin to systematically explore the wider world of story in organization. Classifying stories by genre is just one way to begin to identify the vast number of stories that exist in any organization and begin to understand the rich resource that they can be.

Part III distills and describes four major themes (principles) and attendant behaviors (practices) that were repeated in the stories of this particular group of leaders. Both Parts II and III provide exercises in the form of "reflections" that are designed to evoke stories, interpret their meaning and identify their practical application.

Part IV uncovers the practical implications of Parts I, II and III by illustrating how stories and narratives can be used by leaders to make sense and communicate that sense making to others within their organizations. Part IV suggests how leaders can use their own narratives and the narratives of their organizations to enhance the adaptability and creativity of their organizations and successfully guide their organizations into a prosperous future.

A STORY: *THE STUDY THAT ALMOST FLOPPED*

The following story is offered in lieu of a traditional introduction to begin the examination of how story works in organizations.

I arrived at the Senior Enlisted Academy ready to begin observing, recording and interpreting stories. I was assigned to a private, well-appointed, conference room conveniently located in the same passageway as the students' classrooms. I had a cordial meeting with the directors of the Academy and they seemed genuinely interested in the project. They introduced me to a morning assembly of all of the students and explained what I was there to do. I posted a sign up sheet for interview sessions on the door of my conference room and went off to observe classes fully expecting to return at the end of the day to find the sign up sheet full, but I returned to find the sign up sheet empty.

I rationalized that it was only my first day at the Senior Enlisted Academy and that surely there would be plenty of volunteers on the sign up sheet the next day. When I returned the next morning the sign up sheet was still blank. I continued to sit in on classes observing how the instructors and students used story in their lessons and discussions. Each instructor introduced me to their class and explained why I was there. I made an effort to chat with students during class breaks. The students I chatted with were cordial and seemed interested in the topic. After the second day of observing classes I returned to my conference room to find the volunteer sign up sheet with one name written on it and that name had been crossed off.

I was about to call it quits and catch the next flight home when Master Chief Tom, the curriculum director at the Senior Enlisted Academy, walked by and noticed me slumped at the conference table. He knocked lightly, poked his head in and said, "How's it going?" to which I replied, "Not so good, no one has signed up for my interview sessions, maybe this wasn't such a hot idea." Master Chief Tom raised his eyebrows and replied, "Really?" He went over to scrutinize the blank sign up sheet posted on the door and said, "Hmmm, we probably shouldn't have told them that you were a commander." I replied, "But I'm not in the Navy any more. Do you think I'm intimidating? I have the utmost respect for these guys and gals, that's why I am here. Chiefs are the best storytellers. They have much better stories than officers. They keep this organization going." Master Chief Tom chuckled and said, "Look, I was about to call it a day, but I've got some time right now if you want to interview me." I jumped at the chance to get at least one interview, so I spent the next 90 minutes talking about my project and sharing stories with Master Chief Tom. At the end of the conversation he said "I think we need to talk to the group again tomorrow morning about what you are doing here." When I arrived at 8:00 AM the next morning the sign up sheet on the door was full.

Analysis: *The Study That Almost Flopped*

I do not know exactly what happened between 6:30 PM and 8:00 AM, but it is my guess that through a conversation—through an emerging narrative—a bond was created between me, and Master Chief Tom. Through our conversation he was able to form a better idea of who I was and what I was doing—he made sense of it. He was able to intercede for me with the rest of the group and helped them make sense of what I was doing.

No doubt, a narrative surrounding what I was doing at the Senior Enlisted Academy started with the students the day I arrived. It was percolating in conversations between members of the group in the coffee lounge and during class breaks. The group was coalescing around their interpretation of what I was doing there and they were wary; they were trying to make sense of my interest in "sea stories." The fact that I had been in the Navy perhaps helped me get my foot in the door, but fundamentally I was an outsider with the group. I was not a member of the brotherhood of Chiefs and never had been. Master Chief Tom, perhaps out of genuine interest mixed with a touch of compassion, listened to my story, shared some of his stories with me and a soul-to-soul connection was created. Although Master Chief Tom probably encouraged—perhaps even strongly encouraged—the group to give me a chance, it is difficult to force someone to tell their story unless they are willing to do so. I was not a Chief Petty Officer and never would be, but for a while the group allowed me to have a glimpse into their lives. A happy ending—or in this case a happy beginning—for a project that almost flopped.

PART I

THE NEXUS OF LEADERSHIP, ORGANIZATION AND STORY

'Thou shalt not' is soon forgotten, but 'Once upon a time' lasts forever.

Philip Pullman

Chapter 1

INTRODUCTION

Think about a speech, a conference, a presentation, or a class, you recently attended. What do you still remember about that experience? Do you remember the bullet points in a power point presentation, or do you remember an anecdote the presenter told to illustrate a point? Or perhaps the discussion and real world examples shared during the question and answer session are what stand out in your mind? Perhaps you remember a conversation you had with another attendee on a coffee break? Were you better able to understand a concept presented after engaging in a discourse with others about it? If you can recall stories, anecdotes and conversations better than bullet points flashed on a screen you are not alone. Most humans make sense of what happens in the world around them, including knowledge they are presented with, through narrative interpretation. Story and narrative

---

are such a natural and ubiquitous way of processing information that—like breathing—we rarely recognize "storying" when we are doing it.

The following story was told by Master Chief Tom, the curriculum director at the Senior Enlisted Academy. It describes an incident from early in his Navy career. It is presented here to begin the discussion of the nexus between story and leadership in organizations.

PAGE 10'S

It was 1981. I had just checked into my first duty station, the USS Forrestal [an aircraft carrier]. I was fresh out of "A" School [classroom training in a technical specialty]. I was young, energetic and scared to death. I knew how to get from my rack to the Personnel Office and to chow and that was it. I was afraid that if I went anywhere else on the ship I would get lost. I was working for this Master Chief—Master Chief Bell. I was just awed by him from the start. I had the opportunity to observe up close how much power and influence a Master Chief had—he was like a god—and I just thought, ya know, that's what I want to be someday. So one of my first days there, he assigns me a pile of Page 10's [service record entries] to work on. I really wanted to impress him. So I spent all morning working on those page 10's and they were perfect. I had them all ready for his signature. He was at chow, so I took those Page 10's and put them in a neat stack on his desk and then I went to chow, feeling really good. And when I came back from chow the stack of page 10's was back on my desk all nicely signed. And I was horrified because I didn't know what to do with them. Master Chief Bell obviously expected me to do something with them but I didn't know what. I had never gotten anything back before. They just taught us how to prepare them in A School, not what to do with them after that. So I was just sittin' there staring at that stack of page 10's—not knowing what to do—and finally Master Chief Bell comes out of his office and says to me, "You don't have a clue what to do with those Page 10's now do you?" and I says "No, Master Chief, I am sorry to say that I don't. They didn't teach us that in A School." And he grumbles, "God damn A school." Then he sighs and says "Well, come on shipmate, I'll show you what to do with them since god damn A School did such a piss poor job of training you." And he spent the next hour going through the training manual with me and showing me how to break them down and put them in the service records and all that jazz.

Master Chief Tom went on to tell several more stories about his mentor Master Chief Bell, and in the discussion that followed the telling of this story he said, "Although I only worked for him for a year, and that was almost 30 years ago, to this day when I am faced with a problem I often ask myself, 'What would Larry Bell do?'"

Analysis: Page 10's

On the face of it, *Page 10's* is not an extraordinary story. We all have stories like it. Stories that describe uncomfortable situations that we somehow managed to survive. Stories that reflect how we have learned and grown. If we are lucky we have had mentors like Master Chief Bell who through their example—even in a short period of time—gave us a playbook of behavioral maps that we can reference and adapt to problems we are facing today. Of course, we may have as many, or more, stories about the bosses we loathed as liked, but whether we liked or loathed the people we worked for, and with, it is what we took away from those experiences that for better or for worse formed who we are today. And that is what makes the story *Page 10's* worth exploring here. Something happened in the short period of time that Tom worked for Master Chief Bell that imprinted his character and is still with him 30 years later. *Page 10's* is a story about leadership and followership. It is a story about challenge and character. Tom did not say that he read a book about leadership 30 years ago and memorized principles that he applies to the challenges he faces today—most people simply do not store information in their brains that way. He told a story about someone he admired—someone who formed him. As we listen to that story we get a glimpse into who Master Chief Bell was and who Tom is today.

Some Christians wear wristbands imprinted with "WWJD?" (What Would Jesus Do?). For Christians who have repeatedly heard the stories from Christian Scripture and incorporated the lessons contained in those stories into the way they live, the same principle is at work. Story is the primary way that humans process and store abstract knowledge and information connecting it to other information they have already processed.

What is interesting about *Page 10's* is the implication that 30 years after he worked with Master Chief Bell, Tom is still using the behavioral maps—principles and practices—that he incorporated into his character to guide his actions, and he can articulate examples of the qualities he admired in Master

Chief Bell. Furthermore, he can adapt those qualities to problems he is facing today. *Page 10's* is included here as an example because it illustrates how we think and learn as human beings. We think and learn through narrative interpretation—through story—and that is the primary way we communicate meaning to others. We interpret what is happening around us through narrative and communicate that interpretation through narrative to others.

STORY AND LEADERSHIP

Show me an excellent leader and I will show you an excellent raconteur.

Peters and Waterman

Peters and Waterman in their seminal work on organizational effectiveness—*In Search of Excellence*—said, "Show me an excellent leader and I will show you an excellent raconteur." The connection between storytelling and effective leadership has been known for sometime. Abraham Lincoln, an indisputably great leader, was a masterful storyteller. He purposefully used stories to teach, persuade, motivate, entertain, release tension, repair frayed nerves and salve wounds. He sprinkled his conversations with homespun anecdotes and witty stories whether he was talking to heads of state or farmers in rural taverns. Everyone who came in contact with Lincoln heard him repeat a yarn or anecdote of some kind (Phillips, 1992).

The ability to reach into people's hearts, and engage their passion, through story was not unique to Lincoln and is not unique to company CEO's. Excellent leaders can be found at all levels of an organization and a trait that they all share is that they are excellent communicators. They have the ability to engage and inspire those around them by making knowledge relevant and they do so by telling a story—weaving a narrative that makes sense to others.

Stephen Denning, a former World Bank executive and a prolific writer in recent years on the subject of narrative and leadership, described an epiphany he had in the late 1990s as the program director for knowledge management at the World Bank,

I stumbled onto the power of storytelling. Despite a career of scoffing at such touchy-feely stuff—like most business executives I knew that analytical was good and anecdotal was bad—I changed my thinking because I'd seen stories

help galvanize an organization around a defined business goal (Denning, 2004: 122).

In his books, Denning includes examples of stories that: spark action, transmit values, foster collaboration, tame the grapevine, share knowledge and lead people into the future. Denning contends that organization members and especially leaders can use the art of storytelling to drive strategic change, and leaders who ignore the power and influence of story will likely suffer the consequences (Denning, 2004, 2005).

Like Lincoln, and Master Chief Tom, every leader should have a repertoire of stories at their disposal that they can use to teach a lesson, illustrate a point or make an abstract concept come alive. Every leader should know and be able to articulate the stories that have made them who they are. Every leader should be aware of the stories that are circulating in their organization because, for better or worse, the stories that a leader tells, and the stories that circulate in an organization, will be what the people inside and outside of the organization remember, and those stories will remain with the organization, influencing attitudes and behavior long after the leader is gone.

STORY IN ORGANIZATION

There have been great societies that did not use the wheel, but there have been no societies that did not tell stories.

Ursula K. LeGuin

"So there we were..." or the saltier introduction, "And this is no shit..." are phrases often used in the Navy to launch into a narrative that is known to every mariner from seaman to admiral as a "sea story." Sea stories are one type of narrative commonly used in the Navy and there are similar types of stories in other organizations. Stories may emerge fully formed in speeches or presentations and they constantly pop in discussions, casual conversations and meetings. All organizations have their stories. As the example of *Page 10's* illustrates, stories define who we are and where we have come from. Stories help us cope with life's challenges and create maps for how to proceed into the future. Yet people continue to apologize for their use of story in organizations with phrases like "it is only a story," and very little systematic research has been done to shed light on how story and narrative really works in organizations, which is unfortunate because stories open valuable windows into the emotional, political and symbolic lives of organizations in uniquely illuminating ways.

There have been few large scale systematic studies of the how story in organization works, but since the latter part of the 20th century, there has been growing recognition that conscious use of narrative and story is of value as organizations grapple with challenges regarding how to communicate and manage change. The lack of systematic study of the role that story and narrative plays in organizations, or the lives of individuals, does not suggest that storytelling is not valued in organizations. Humans have been telling stories since they started communicating through language and everyone enjoys a good story. The importance of story in the evolution of humans is evident in early attempts to communicate meaning in mediums as diverse as cave paintings, hieroglyphics and ancient biblical texts. It is our ability to use narrative and story to pass on our knowledge to others that has separated us from other animals and led to our understanding of the universe. (A more detailed history of narrative and story from the ancient Greeks to the present is included at Appendix B.)

Although there have been few large, systematic studies of story in organizations there was a tangential piece of a study that suggested story may play a significant role in how work gets done in the Navy. In a study of organizations that perform inherently dangerous and highly technical tasks—that included Navy aircraft carriers—Rochlin *et al.* (1987) suggested that story plays an important role in the training that takes place onboard Navy ships. They found that flight operations on board aircraft carriers were "the most extreme in the least stable environment" yet, from a safety perspective, aircraft carrier crews performed better than other high risk organizations, despite the fact that the crews are largely young and inexperienced, and the management team turns over every 24 to 36 months. The researchers suggested that story is one of the primary ways knowledge is transmitted on board these ships and inferred that story played a significant role in the excellent safety record. This tangential piece of a research finding suggests that stories may play a highly significant communication role in the Navy (Rochlin, La Porte & Roberts, 1987).

STORY AND NARRATIVE DEFINED

Before proceeding into any further discussion of "story" and "narrative" it is important to define what is meant by these terms as they are used throughout this book and it is important to note that others may define story and narrative differently. The term "story" as it is used here is for the most part synonymous with "anecdote," but it is more than anecdote because it also may include the discourse surrounding an anecdote that underscores its meaning or invites further interpretation. A "story" in its common usage usually presents a problem—dissonance of some sort—that the characters in the story in addition to the audience and perhaps the teller, must try to resolve, so that the story fits with the characters' view of reality and how the world works. Through characters, settings, motives, struggles, conflicts and plot, something is revealed as the story unfolds—an idea, a moral, a message—for the audience to take away. Stories often involve making the ordinary seem extraordinary or the extraordinary seem ordinary. A narrative may include an anecdote, a story, or stories, in addition to the discourse that puts the story into a context and connects it to the thoughts that come before and after. But a narrative may not be a fully formed story with a beginning, middle and end. Narratives usually follow a thematic thread but they are often evolving and can be fragmented. The meaning of some stories is crystal clear but often stories need narrative interpretation to uncover their meaning and relevance. Stories make up a larger narrative that reflects who and what we are.

In the later part of the 20th century and the beginning of the 21st century, the term "narrative" gained currency and legitimacy in broadcast and print media. The media started to regularly refer to "narrative," for example, "The president failed to present a compelling narrative," the debates about economic crisis involved "competing narratives" and the leadership needed to present a "coherent narrative." Through narration the teller describes a plot, that is to say they describe attempting to make sense of something that does not make sense for one reason or another. Sometimes the teller has already "made sense" and is communicating that "sense making" to others. But just as likely the narrator is in search of a coherent narrative that will help them make sense. Sense making, and the weaving of a narrative, is an ongoing individual and group process.

Through narrative the teller may be testing a plot to see if it fits with other stories, narratives, or knowledge, the teller, or the listener, holds. Both narratives and stories may change as more information is collected and as humans interpret events to fit with information they have already collected or interpreted. For example, the information contained in a report of declining or increasing unemployment figures must be interpreted to be useful and have meaning; once interpreted it will become part of a narrative made up of stories and information threads that create meaning, including explanations of cause and effect, within a broader context.

Both stories and narratives are a subjective interpretation of events and therefore may not be strictly factual; they reflect life as the individual sees it and, like the Buddhist parable of *The Blind Men and the Elephant* (See Sidebar: *The Blind Men and the Elephant*), different individuals will rarely, if ever, interpret reality in exactly the same way. That being said, organizational stories are rarely total fabrications; they are usually a rich reflection of the organizational culture that spawned them.

Stories give meaning and structure to life events. Narratives may be complete, or a work in progress, but neither stories, nor narratives, are static—at least not for long—they are constantly shifting and changing through interpretation, modification and reinterpretation. Stories and narratives increase understanding and bring people together in a common perspective. Ultimately, all stories and narratives are human attempts to cope with the daily business of living in a complex universe, whether they increase understanding or not.

Models adopted from natural systems are especially well suited to explaining how story and narrative works in organizations. Like sticks tossed into a rushing stream stories rarely follow a linear path. They may follow a similar trajectory but they may be thrown off course by the narrative equivalent of a gust of wind, the wake of a passing boat, or a dog playing fetch. The child's party game "telephone" is an amusing illustration of how difficult it is to force story, or narrative, along a linear path. Some stories move quickly through an organization while others seem to die out, or perhaps disappear and reappear at a later date. Stories are interpreted and re-interpreted over time, adapted to suit the purpose at hand and are sometimes stripped to their essence. Sometimes all it takes is a single phrase to communicate the essence of a story, such as the admonition "Don't cry wolf," or the often used phrase "You know the rest of the story."

A NEW WAY OF LOOKING AT THE WORLD

The way we see the world determines how we manage it.

W. Edwards Deming

W. Edwards Deming, father of the "Total Quality" movement that was embraced by the Navy and many large "business" organizations in the latter part of the 20th century, said "The way we see the world determines how we manage it," (Goerner, 1999). We all view the world through our own personal lenses—developed from the stories and narratives that make us who we are. Understanding our personal lenses and the lenses of the groups we are a part of can help us understand where we need to change, or adjust our view, to adapt and thrive. Every once in a while an old way of looking at the world fails—it ceases to explain what is happening and why—an indication that a new lens is needed. It is the adept, self aware, leader who recognizes when a new lens is needed. The clues to why old ways are falling short is in the narratives—the narratives surrounding the interpretation of the data and the narratives that define reality for the people inside and outside of the organization.

NASA's Challenger and Columbia accidents are a case in point. NASA has been publicly cited (CAIB, 2003) as an organization that has, at times, disregarded the importance of organizational culture and narrative, resulting in the perpetuation of flawed decision making that contributed to both the Challenger and Columbia space shuttle accidents. The Columbia Accident Investigation Board reported that the organization was not responsive to criticism following the Challenger accident due to an ingrained bureaucratic culture that became rigidly defensive when faced with criticism. The organization saw data that supported a narrative interpretation it already held as truth, and was reluctant to accept data that did not fit with that interpretation.

A narrative that brings together seemingly isolated events into a coherent sequence can help an organization make retrospective sense of seemingly disconnected pieces of data. A narrative interpretation can then be used to illuminate causes in view of all the data and thus facilitate organizational change to perhaps reduce the repetition of similar mistakes in the future. Only after looking at the broader world in which the facts are couched and being open to the possibility that the facts can be interpreted in a variety of ways can we begin to hope to find the truth or perhaps many truths.

The following story—*Smith's Cranial*—is included here as an example of how stories travel in a non-linear fashion through organizations connecting people and serving different purposes at different times. No doubt, there are similar stories in other organizations. *Smith's Cranial* illustrates how story develops a life of its own.

SMITH'S CRANIAL

The Arabian Sea...zero one hundred hours...no moon...less than ideal conditions for night flight operations. An aircraft carrier—97,000 tons of floating city—pitches and rolls in the inky blackness. A phalanx of jets returning from flight operations hit the 300 foot landing zone on deck every 30 to 45 seconds.

In the passageway directly below the flight deck Seaman Apprentice Harting (age 19) works with Chief Reiter (age 33) on an electrical panel. A dented flight deck cranial [safety helmet] with the name "Smith" neatly stenciled across the back, hangs by the ladder to the flight deck. The young seaman watches the cranial shudder each time a jet slams into the deck. The electrical panel repaired, the young sailor asks: "Hey Chief, who is Smith and why is his cranial hanging there?" The Chief, chewing on a toothpick, glances over his shoulder at the cranial as he secures the electrical panel, and says, "Lemme tell you, and this is no shit... We had been in the Indian Ocean for ninety days and word was passed that a COD [aircraft with cargo] was inbound with mail and a load of ice cream. Smith was part of the supply crew. We had other fixed wing aircraft inbound so we had to get the COD unloaded and off the deck before the rest of the planes came in. Young Smith, in a sweat to get his ice cream, or a letter from his girl friend, or whatever, almost went on deck without his cranial. The Master Chief caught him up short just about where you are standing and says, 'Hey Smith, where do you think you are going without your cranial? I don't want to be writing a letter to your mother after your brains—what little you have—are spilled all over the deck. I got enough paper work to do.' So Smith double times it back to get his cranial. And lucky he did, in the sweat to get the COD unloaded Smith took a glancing prop strike to the head. If he didn't have his cranial on his brains would have been spilled all over the deck and some young sailor—just like you shipmate—would have had to clean them up. So the Master Chief convinced the skipper that we should hang Smith's cranial up there as a reminder."

Six months after Seaman Harting asked about the cranial hanging in the passageway he was painting the bulkhead in the same passageway with Seaman Apprentice Schiff (age 18), who reported on board last week. Seaman Apprentice Schiff said to Seaman Harting, "Who the f--- is Smith and what is his cranial doing there?" Seaman Harting replied, "Lemme tell ya, and this is no shit...last deployment Smith was assigned to a crew that was unloading a COD loaded with ice cream. They were in a sweat to get the COD unloaded because they had fixed wing coming in and they needed to get the COD off the deck. Smith took a strike to the head. The first one nicked him but the second one took his head right off." Seaman Apprentice Schiff said, "No shit, were you there?" Seaman Harting replied, "Naah, but a buddy of mine was. He had to clean up the blood and brains and shit that were spilled all over the deck." Seaman Apprentice Schiff said, "No shit?" Seaman Harting, "No shit, and they ended up giving the dude a medal." Seaman Schiff, "Did they get the f---ing ice cream unloaded?" Seaman Harting snorted, "I guess so. That's probably why they gave the dude a medal." Seaman Schiff chortled, "They should have given him a f---ing purple heart!"

Postscript: Ten years later Chief Smith had occasion to meet Chief Harting with his head in tact and that is yet another story.

Analysis: Smith's Cranial

Smith's Cranial is not a single story, it is a nesting of stories within a larger narrative. There is the story of the cranial told by Chief Reiter, it contained a sequence of events and a problem—a plot—and it was told to teach a lesson. There is the story of the cranial told by Seaman Apprentice Shiff, it contains a different sequence and plot, and it was told to entertain and elevate the status of the teller—an immature sailor—rather than teach a lesson. There is the story that the cranial—a symbol, a cultural artifact—represents to all who pass it each day, each interpreting it in their own way. And finally there is the story as it was presented here, couched in a narrative, as an example of a sea story, its purpose was to illuminate the dynamic nature and complexity of story as it moves through an organization, adapting to new situations and branching out in a sometimes twisted web.

In the following chapters the reader will see how Navy Chiefs, as individuals and as a group, use their stories and narratives to adapt to constantly changing environments. An examination of the stories of Chiefs reveals how they use story and narrative to: make sense, train, educate, mentor, socialize, persuade, lead, resolve dissonance and facilitate change. The findings have implications that reach well beyond the Navy as more organizations search for optimal ways to function and thrive in a fast paced, ever changing world.

CHAPTER SUMMARY

The leadership principles and practices of Navy Chief Petty Officers (Chiefs) have never been systematically studied, yet Chiefs have been very effectively accomplishing the bulk of the work that supports the war-fighting mission of the Navy for decades. An exploration of the stories and more expansive narratives of Navy Chiefs uncovers how they collaborate with each other, partner with Commissioned Officers and motivate young sailors to accomplish the often unglamorous, behind the scenes, tasks that need to get done.

This book also presents an opportunity for a philosophical paradigm shift with respect to the way leaders and managers view the world and how organizations function, suggesting that biological metaphors rather than machine metaphors better explain how human organizations work. Organizational stories and narratives reflect the living, vital essence of organizations. A cycle of narrative meaning develops in organizations as people use stories to make sense of the past, cope with the present, and proceed into the future. This cycle of narrative meaning repeats itself as the present becomes the past, and the future becomes the present, with stories ebbing and flowing between sense making, defining reality, and providing maps for how to proceed into the future. Stories travel in a non-linear fashion through organizations, facilitating connections between people and creating meaning.

The following chapters explore the stories and narratives of Navy Chiefs providing a model for how others can connect with their own stories and the stories of their organizations to successfully navigate into the future. Through the stories of Navy Chiefs, the reader will be able to see how stories bridge the gap between sensed information, abstract knowledge and technical information, and how that knowledge fits into a complex whole that is constantly shifting and changing as organizations move through space and time.

Sidebar: *The Blind Men and the Elephant*
by John Godfrey Saxe (1816-1887)

It was six men of Indostan
　　To learning much inclined,
Who went to see the Elephant
　　(Though all of them were blind),
That each by observation
　　Might satisfy his mind.
The First approached the Elephant,
　　And happening to fall
Against his broad and sturdy side,
　　At once began to bawl:
"God bless me! But the Elephant
　　Is very like a wall!"
The Second, feeling of the tusk,
　　Cried, "Ho! What have we here
So very round and smooth and sharp?
　　To me 'tis mighty clear
This wonder of an Elephant
　　Is very like a spear!"
The Third approached the animal,
　　And happening to take
The squirming trunk within his hands,
　　Thus boldly up and spake:
"I see," quoth he, "the Elephant
　　Is very like a snake!"
The Fourth reached out an eager hand,
　　And felt about the knee.
"What most this wondrous beast is like
　　Is mighty plain," quoth he;
"'Tis clean enough the Elephant
　　Is very like a tree!"
The Fifth, who chanced to touch the ear,
　　Said: "E'en the blindest man
Can tell what this resembles most;
　　Deny the fact who can
This marvel of an Elephant
　　Is very like a fan!"
The Sixth no sooner had begun

About the beast to grope,
Than, seizing on the swinging tail
That fell within his scope,
"I see," quoth he, "the Elephant
Is very like a rope!"
And so these men of Indostan
Disputed loud and long,
Each in his own opinion
Exceeding stiff and strong,
Though each was partly in the right,

And all were in the wrong!

Moral:

So oft in theologic wars,
The disputants, I ween,
Rail on in utter ignorance
Of what each other mean,
And prate about an Elephant
Not one of them has seen!

Chapter 2

THE SETTING AND THE CHARACTERS

Every day at sea is a good day.

Master Chief Cal, Master Chief Tom, Master Chief Jeffers

This book focuses on the keepers of Navy culture—Chief Petty Officers (Chiefs). Navy Chiefs, as the elders with the most practical experience in an organization of young people—most members are in their 20s and only eight percent of Navy personnel are over 40—preserve and pass on Navy culture in the form of stories, quips, jokes and other symbols.

Every human organization has a unique culture that encompasses the informal values and norms of the culture and influences the attitudes and behaviors of the individuals within the organization (George & Jones, 1996).

Organizational culture gives the people within an organization a way of behaving toward each other and the world outside of the organization. Effective leaders shape, guide and invigorate organizational culture, and they have been doing so since the days that humans organized themselves into tribes and clans to survive and prosper.

Like any organizational culture, it is hard to truly understand the culture of the Navy unless you have been a part of it, but there are many similarities with other organizations. Anthropologists create descriptive narratives of the cultures they study in an attempt to paint a picture for others. A descriptive narrative follows, so that those who are unfamiliar with the Navy will have a better understanding of the characters and settings that are the backdrop of the stories in this book. Navy insiders might not need a description—they know these characters and have lived in this world—but the description included here is one person's interpretation replete with biases and opinion. Hopefully even Navy insiders will find the following description of their world illuminating.

THE SETTINGS: A LEADERSHIP ACADEMY AND A WARSHIP AT SEA

The stories and narratives of the Navy Chiefs that are included here, were collected at two geographically separate sites in the United States with different missions and functions: (1) The Senior Enlisted Academy, a leadership academy that trains Navy Chiefs and non-Commissioned Officers from other branches of the Military who are transitioning into positions of increased authority and responsibility and, (2) a Navy aircraft carrier operating at sea.

The Senior Enlisted Academy

The Senior Enlisted Academy offers a six-week program designed to develop the leadership ability of Navy Chiefs. There are approximately 65 students enrolled in each class. The class is divided into small study groups to allow for the sharing of experiences and ideas. Students must apply and be accepted into the program. Priority is given to those who are transferring into positions of expanded leadership and management responsibilities in the Navy. Some senior enlisted members of the other branches of the United States Military— and a few senior enlisted members of allied countries—are enrolled in each class. The courses are taught primarily by Navy Master Chiefs who have been recognized for their exceptional leadership ability, and chosen through a highly selective process to be instructors and directors at the Senior Enlisted Academy. All students live on site; they engage in physical training and team building exercises as well as academic training while there.

The Ship

The other site visited was a Navy warship—an aircraft carrier. The primary mission of an aircraft carrier is launching and recovering aircraft. Aircraft carriers are often aptly described as "floating cities" replete with all the operational and functional components of a city where people must work and live—they are the largest warships in the world. There are approximately 70 aircraft on board an aircraft carrier along with all the ancillary services it takes to support them, including weapons systems for defense of the ship and its embarked aircraft. The flight deck is over three football fields in length and covers 4.5 acres. Aircraft carriers, such as the one described here, are nuclear powered so they can conceivably remain underway without needing to refuel for decades. However the aircraft, and the vehicles used to move them and other equipment around the ship, are all powered by jet fuel that must be replenished at regular intervals. The carrier is alive with activity around the clock. When the carrier is at sea the crew routinely works in 12-hour shifts—12 hours on and 12 hours off. There are television lounges where crewmembers can watch movies or television programs 24 hours a day. There are gyms, a library, a chapel, 24 hour a day food service, laundries, and even an internet café. Up to 5,000 sailors live and work on an aircraft carrier when it is at sea. Approximately 300 of them are Chiefs. Approximately one sixth of the crew is female. Women have relatively recently—since 1995—been assigned as crewmembers on aircraft carriers. Prior to that time the "Combat Exclusion Law" (Section 6015 of U.S. Code 10) prohibited women from serving on warships. The numbers of women serving at sea has gradually increased since the legislation was changed.

To an outsider, the activities on the flight deck of an aircraft carrier during flight operations might appear chaotic. There are jets slamming into the deck and blasting off, cables whirring, wind blowing, jet engines revving up and "spooling" down, small tractors crisscrossing the deck with large aircraft in tow, and people scurrying about everywhere talking into radios and waving hand signals at one another. The noise level is so intense that everyone working on deck is required to wear ear protection, and no one who is not supposed to be there goes on deck during flight operations. But Senior Chief Tim, the Chief Petty Officer in charge of moving the aircraft around the flight deck and to the hangar bay below, described flight deck operations as "a carefully choreographed ballet." Every other department on the aircraft carrier is focused on supporting flight deck operations—the primary mission of the ship.

THE CHARACTERS: CHIEF PETTY OFFICERS

Your entire way of life has now been changed. More will be expected of you; more will be demanded of you. Not because you are an E-7, but because you are now a Chief Petty Officer.

The Chief's Creed

Navy Chiefs are often described as "deck plate leaders," they operate in the middle management layer of the Navy where the proverbial "rubber meets the road." They do not fly planes, command ships, or formulate tactical battle plans. They are responsible for getting the often dirty, grueling, sometimes tedious and sometimes terrifying behind the scenes work done in support of the Navy's war fighting mission. They must motivate and train inexperienced sailors—the young men and women who comprise the bulk of the Navy's work force—to commit their hearts, minds and bodies to accomplishing these unglamorous tasks, often when they least want to work. And Chiefs must coach, and collaborate with, young Commissioned Officers—most of whom are right out of college and ostensibly their superiors in the chain of command—to manage all aspects of the workload assigned to a division of a few up to over 100 people with a highly specialized mission.

This book focuses on the stories of Navy Chiefs because, as a group, they are superb storytellers and they also have a well-deserved reputation for their ability to get work done. Two phrases often heard in the Navy are "If you need an answer, ask the Chief" and "If you want to get a job done, ask the Chief." Add to these, "If you want to hear a good story, ask the Chief." This book includes the personal stories of Chiefs about their experiences and stories told by others who have been inspired and influenced by them.

All Chiefs have a minimum of eight years of full time military service. They are "non-commissioned officers," that is to say they joined the service at the lowest possible pay grade—often right out of high school—and have risen through the ranks to fill middle management and leadership positions. Enlisted members are promoted to the rank of Chief as a result of their competence in a technical specialty, and their demonstrated leadership ability. Enlisted members and "non-commissioned officers" enter into contracts to serve for a set period of time, usually two, four, or six years. After their initial contract is fulfilled they may reapply for an extension or re-enlistment on their contract for another set period of time.

In contrast, "Commissioned Officers"—such as, Ensigns, Lieutenants, Commanders, Captains—usually enter full time military service after receiving a college degree. In the Navy Commissioned Officers may be graduates of the United States Naval Academy or they may be graduates of other four-year colleges. Commissioned Officers initially serve for a set period of time, usually five or six years, and may incur further commitment as a pay back for advanced education such as graduate school or flight training. Commissioned Officers stay on active duty service until they resign or are asked to resign. A Chief Petty Officer could apply for a commissioning program and become a commissioned officer. And while not impossible it is highly unlikely that a Commissioned Officer could ever become a Chief Petty Officer.

Young junior officers may have the academic accreditation that comes with formal education, and they may have held leadership positions, for example as student leaders at the Naval Academy or on college campuses, but unless they were previously enlisted, they lack the practical experience that most Chiefs possess. Most divisions—the smallest unit of organization in the Navy—are assigned both a junior officer, in the rank Ensign or Lieutenant, and a Chief Petty Officer, to lead and manage the operations of the division. While technically superior in the chain of command, it is the wise junior officer who teams up with their division Chief to learn the ropes of the organization. And it is the wise Chief who cultivates a strong partnership with his or her Division Officer grooming the young junior officer to liaise with superiors in the chain of command to obtain authority and material support to accomplish the division's mission.

Individuals selected for Chief Petty Officer are subjected to an intense initiation period through which they are indoctrinated, trained and tested by their fellow Chiefs before they are accepted into the ranks of the "Chiefs' Mess." The Chiefs' Mess is the physical location onboard a ship where the Chiefs eat their meals, hold meetings and relax, but the term has also come to represent a strong, quasi-official, fraternal organization that reaches across the entire middle management stratum of the Navy. The Chiefs' Mess is an accepted organization that self-organized decades ago and evolved over time to fill a need. The Chiefs' Mess plays a crucial role in the Navy today. It might not occupy a box on any official organization chart, yet it is often referred to as if it is officially sanctioned and tasked to accomplish aspects of the organization's mission. The networked and self-organized aspects of the Chiefs' Mess will be discussed in greater detail later.

The individuals in the Chiefs' Mess represent a wide range of technical specialties as well as a wealth of diverse, practical experience. There are three levels of Chief Petty Officers: 1) Chiefs (pay grade E-7), 2) Senior Chiefs (pay grade E-8) and, 3) Master Chiefs (the most senior, pay grade E-9). Each organization usually has a Master Chief Petty Officer of the Command (CMC) who reports directly to the Commanding Officer and there is a Master Chief (MCPON) of the Navy who reports directly to the Chief of Naval Operations. (See Sidebar: CPO Creed).

The Navy is an organization of young people with high turnover (the entire crew of a ship turns over on average every three to four years). Navy members are either promoted and advance in their technical specialty or their contract is not renewed and they are asked to leave. An enlisted person achieving the rank of Master Chief Petty Officer may stay in the Navy for 30 years. A few Master Chiefs remain on active duty longer than 30 years but they are an exception. Therefore, a Chief Petty Officer in his forties or beyond is one of the respected elders in the organization. All of the stories presented and discussed here are the stories of Chiefs, Senior Chiefs and Master Chiefs.

The Chiefs whose stories are presented here were all volunteers. Group and individual interviews were conducted to evoke stories and narratives and the volunteers were shadowed on the job to observe their use of story and narrative in the course of a normal workday. All of the Chiefs interviewed, and observed on the job, had between nine and 30 years of service in the military. Their race and ethnicity varied, as did their technical specialties. (A snapshot of participants is included at Appendix C: Table 1 and Table 2.)

Why examine the stories of Chief Petty Officers?

A theory is a hunch about something we intuitively know. As I mentioned earlier, I had a hunch—a theory—about Chiefs. I knew from personal experience that Chiefs are excellent storytellers, and I had a hunch that Chiefs as a group do not work like any other group in the Navy. I also had a hunch that there was a correlation between storytelling and the ability of Chiefs, as a group, to work through and with others to get the job done.

Chiefs are well known for their unparalleled ability to network. Indeed, Chiefs' ability to network across organizational lines, as well as up and down their own chain of command through exercising the art of "cumshaw"—a Navy term for acquiring what is needed through unorthodox methods—is legend. I had a hunch that the ability to weave a narrative played into Chiefs' ability to connect with others to get what they needed to accomplish the mission. Indeed, the observations of Chiefs at work and the stories collected from them confirmed that not only are Chiefs exceptionally adept at networking, they also often function as key nodes in the networks that they are a part of, facilitating or blocking the flow of energy and resources in the organization. Story is the spark and the current that propels the network of the Chiefs' Mess.

Suggesting that work gets done in a networked, web-like, fashion rather than a linear fashion is a radical notion for a classically organized command and control type organization like the Navy. Observations from natural systems illustrate that organizations are structures for energy flow, and natural systems are usually webbed rather than linear. Energy in human organizations flows through interaction and communication within the organization, as well as communication with the environment surrounding the organization. Narratives are the conduits of thoughts and ideas that course through communications networks inside and outside of human organizations.

Size can be the enemy of speed when it comes to energy flow; the key to successful adaptation is to stay well linked so that energy flows freely (Barabasi, 2002). The Navy is a huge organization yet creative problem solving takes place regularly in the Navy because of the flow of ideas that takes place in the multitude of networked groups that comprise the organization as a whole. Creative problem solving takes place when networked groups tap into their collective knowledge base and stay linked through the mindful use and awareness of the narratives and stories swirling within, and outside of, the organization. The networking of Chief Petty Officers across and within organizational lines helps keep the organization nimble. The Chiefs' Mess when functioning at its best is like a biological system tapping into the energy of its network.

There was evidence in the observations of Chiefs at work and in their stories that the Chiefs' Mess is more of a web than a hierarchy, and more collaborative than competitive as compared to the "Wardroom"—the collective group of Commissioned Officers in a Navy organization—or surface warfare officers as a group, or naval aviators as a group. And in an interesting adaptation of a classically structured command and control organization, not only does the Navy hierarchy tolerate the networked behavior of the Chiefs' Mess, it actually seems to encourage it. It is my hunch that this webbed network approach toward accomplishing work—which is more akin to the way natural systems work than the way mechanical systems work—is better suited to the rapidly changing work environment of the 21st century.

There was ample evidence in the Chiefs' stories and narratives of pooled insights brought to the task of problem solving through the Chiefs' Mess. The Chiefs repeatedly referred to getting work done "in the Chiefs' Mess" or "through the Chiefs' Mess." The stories about Charge Books, references to the Chief Petty Officers initiation, and multiple references to taking care of one's brothers or sisters "in the Mess" were evidence of networking that transcended even race and ethnicity which are typically strong identifiers in western cultures and subcultures.

One of the students at the Senior Enlisted Academy, Senior Chief Dee, described the importance of the network of the Chiefs' Mess as follows.

It (the Navy Chiefs' Mess) is a network you can tap into. If you have a question or you need something you start with your own Chiefs' Mess but if you can't find the answer there, or get what you need, you reach out to the rest. For example when I was in Bahrain, the Chiefs' Mess on my ship was small but we had 350 Chiefs in the Mess throughout Bahrain that you could tap into if you needed something. You could send out an e-mail and say "Does anybody have thus and such?" and nine times out of ten one of your brothers or sisters in the Mess would be able to help you out or know someone who could.

Master Chief Cal, the Chief in charge of the enlisted component of largest department on the carrier, said that he gets most of his work done "in the

Mess" on the Ship. Master Chief Cal was referring to the physical space on the Ship where the Chief Petty Officers eat their meals, but he was also referring to the network of the Chiefs' Mess on the Ship and beyond. He explained,

The Chiefs' Mess on a ship represents thousands of hours of experience from different ships at different times. There is rarely a problem that comes up that someone in the Mess has not seen before, or at least seen something similar. And there are lots of different perspectives because of the different technical specialties represented.

Master Chief Cal's references in his stories, and the discussions surrounding them, to the network of the Chiefs' Mess, and the references of other Chief Petty Officers to the collaborative nature of the Chiefs' Mess, were evidence of a strong network that uses narrative and story around the clock as it grapples with challenges.

Master Chief Joe, the Master Chief of the Engineering Department on the Ship, called the Chiefs' Mess "the glue" that holds the Navy together. Many of the other Chiefs reflected similar sentiments. He explained how the Chiefs' Mess is different from the group of Commissioned Officers on the ship—the "Wardroom,"

There really is not much competition among the Chiefs in the Mess because for the most part we all have different technical specialties—we are not so much competing against each other as against other Chiefs off the ship that are in our rating, and even then the competition is not so great because once you make Chief you are it and it is what it is, I mean you have sort of made it, sure you want to make E-8 and E-9, but you really aren't competing with the Chiefs on your ship for that. Maybe I'm wrong, but I sense that there is more of a bond and less competition (in the Chiefs' Mess) than there is in the Wardroom.

Master Chief Tully, one of the aviation logistics specialists on the Ship, cited the tradition of the "charge book" as an example of the wide-ranging network of the Chiefs' Mess. Each newly selected Chief is tasked with creating a charge book during their initiation into the brotherhood of Chiefs. They are "charged" with getting other, more experienced, Chief Petty Officers, to make entries in their charge book. The entries may be in the form of advice, lessons, jokes, anecdotes, drawings and words of wisdom; and the initiates are also sometimes "charged" monetarily for non-compliance with those entries. Master Chief Tully proudly displayed his charge book on his desk. It was a three-inch thick tome that looked like a family bible with its elaborately carved wooden cover and leather straps. Each page of Master Chief Tully's charge book contained leadership narratives and practices. The charge book is an example of a cultural artifact used to encapsulate the values of the organization that have been passed down from generation to generation of Chiefs. Each charge book is living proof that Navy Chiefs are the keepers of a culture that sustains the organization.

CHAPTER SUMMARY

Asampling of the stories and narratives of Chief Petty Officers (Chiefs) at two separate sites—a leadership academy and a warship at sea—are explored in this book. Chiefs are career enlisted men and women in the Navy who have risen through the ranks into middle management positions. They are superb storytellers who have a well-deserved reputation for their ability to network across organizational lines as well as up and down the chain of command to get work done. They are the keepers of a culture that largely determines how the Navy works.

SIDEBAR: THE CHIEF'S CREED

During the course of this day you have been caused to suffer indignities to experience humiliation. This you have accomplished with rare good grace and, therefore, I now believe it fitting to explain to you why this was done. There was no intent, and no desire, to demean you, nor to insult you. Pointless as it may have seemed to you, there was a time-honored and valid reason behind every single deed, every single barb. By experience, by performance and by testing, you have been this day advanced to Chief Petty Officer in the United States Navy, and only in the United States Navy does E-7 carry unique responsibilities, no other armed force throughout the world carries, nor which grants privileges to its enlisted personnel comparable to the privileges and responsibilities you are now bound to observe and are expected to fulfill.

Your entire way of life has now been changed. More will be expected of you; more will be demanded of you. Not because you are an E-7, but because you are now a Chief Petty Officer. You have not merely been promoted one pay grade. You have joined in an exclusive fraternity and, as in all fraternities, you have a responsibility to your brothers even as they have a responsibility to you. Always bear in mind that no other armed force has rate or rank equivalent to that of the United States Navy. Granted, that all armed forces have two classes of service: enlisted and commissioned; however, the United States Navy has the distinction of having four, i.e., Enlisted, Bureau appointed CPO, Bureau appointed Warrant and Commissioned. This is why you can maintain with pride your feeling of superiority once you have attained the position of E-7 in the United States Navy.

These privileges, these responsibilities, do not appear in print. They have no official standing. They cannot be referred to by name, number or file. They exist because for over 200 years the Chiefs before you have freely accepted responsibility beyond call of printed assignment and have, by their actions and performance, commanded the respect of their seniors as well as their juniors. It is now required that you be a fountain of wisdom, the ambassador of good will, the authority in personnel relations as well as technical applications. "Ask the Chief" is a household phrase, both in and out of the Navy.

You are now the "CHIEF"! So this, then is why you were caused to experience these things. You were subjected to humiliation to prove to you that humility is good, a great, a necessary attribute which cannot mar you—in fact, it strengthens you—and, in your future as a Chief Petty Officer, you will be caused to suffer indignities, to experience humiliation far beyond those imposed upon you today. Bear them with the dignity, and with the same good grace which you bore them today! It is our intention to prove these facts to you and to accept you. Your performance today has assured us that you will wear your hat with aplomb, as did your brothers in arms before you. We take a deep, sincere pleasure in clasping your hand and accepting you as a Chief Petty Officer in the Untied States Navy.

PART II

USING STORY GENRE TO BUILD A STORY REPERTOIRE

Those seeking a pre-packaged catalog of stories that they can use to communicate complex ideas and motivate people will not find one here. That is not to say that excellent collections of stories do not exist. Most industries have collections of case studies that are stories with a pre-determined moral that can be effectively plugged into lesson plans and presentations. Classic stories, like pre-packaged jokes and anecdotes, add spice to speeches and presentations, but they need to be part of a larger, emerging narrative that adapts them to the purpose intended, in order to communicate meaning.

This book takes a different approach, providing examples and asking the reader to develop their own repertoire of stories that have unique relevance for them and their organization. This is not an easy process. The stories are within us but we must make an effort to uncover them and recognize how they have made us who we are and how they will influence who we will become if we are to fully tap into the power of story. The exercises in this book provide some structure for creative thought. They will help the reader uncover their personal stories and explore how they can inspire behavior, explain abstract ideas, illuminate truth, or spark soul-to-soul connections.

Identifying personal, or organizational, stories in terms of classical genres—such as "epic," "tragedy," "comedy" and "romance" and combinations of those types—is one way to begin to categorize and evoke the stories that guide human organizations and make us who we are. Stories may have similar themes or plots but they also can have infinite variety and be adapted to diverse and multiple purposes. Classifying stories by genre gives some much-needed structure to the interpretation of stories and how they can be used to help people make sense, cope and thrive.

Chapter 3

EPICS

I had my doubts—but I never let the troops know that—I just kept telling them 'We're gonna ace it.'

Master Chief Arcelo

The following epic story was told by Senior Chief Tim, a self-described "fully pose-able GI Joe with kung fu moves." Senior Chief Tim is a gregarious, muscular, Aviation Boatswain Mate who always seemed to have an audience. He told the following story while standing on the flight deck of the Ship during a lull in flight deck operations.

USS KITTY HAWK MISHAP

So there we were… It was a just another typical day at sea. The weather was good. Flight ops were underway. I was standing right about here on the flight deck with a couple of other guys when all of a sudden—BAM—this F-14 [jet] hit the deck. It didn't sound right—and of course things were happening really fast—but I remember seeing this incredible flash of orange out of the corner of my eye. It took a second to register that the plane was on fire—in fact, the plane had split in two. So I call over my shoulder to the guys I was standing with ,"Let's go!" and I ran over to the connectors for the fire fighting equipment. I could see fuel on the deck. And I go to pass it (the fire fighting equipment) over my shoulder to the guys who I thought were behind me and they weren't there—some people just freeze when something like that happens. Then I noticed the pilots had ejected but they weren't in the water. One of them was in the netting [the safety netting that surrounds the flight deck]. He looked conscious and someone was running toward him so I figured they had him. Then I noticed the other pilot still in his seat pan [the seat of the aircraft that ejects with the pilot] with the chute [parachute] attached—he was not far from where I was. He was in pretty bad shape. He was shaking like he was in shock. His sleeves were rolled up—which is a "no, no"— and there was skin hanging off—it was pretty ugly. By that time a couple of other guys had come over to fight the fire so I ran over to the pilot lying on the deck to try to release him from the seat pan and parachute [the parachute is attached to the seat pan]. I tried to push the buttons to release the seat pan but my fingers are too big, I could not get them in the holes to push the buttons that release the pan. So I am hanging on to the guy and the seat pan and the chute starts to inflate and now we are both being dragged down the deck at a pretty good clip as the chute continues to inflate. At that point I figured I would just hang on to the guy and we would go into the water together. We both had our flotation gear on so I thought at least if I stuck with him I could keep us both afloat. So there we were moving down the deck pretty fast now. My pants were ripped off as I scrapped along the non skid [the deck of the ship that is like heavy grade sand paper] and my bare ass was hanging out and scrapping along the deck. It wasn't pretty. We were real close to the edge and I thought this is it, we're going into the water, when one of the flight crew came running out, reaches into the pilot's flight suit, pulls out the pilot's safety knife and cuts the lines [ropes] on the chute just as we were about to be dragged off the deck. Luckily he knew where the pilots keep their safety knives.

Very few people survive the fall of approximately five stories from the flight deck of an aircraft carrier without sustaining serious injuries or death. Senior Chief Tim risked his life to save a shipmate who was complete stranger. He likely would have been injured and may have been killed if he, and the pilot he was attempting to rescue, had been blown off of the deck of ship. The partially inflated parachute might have slowed their descent into the water, but it just as easily could have become tangled in the superstructure of the ship and functioned like a tether dragging them along in the wake of the fast moving ship or slamming them into the side. If senior Chief Tim had not survived the story would have been a "tragedy" not an "epic".

Senior Chief Tim was a bona fide hero in the eyes of those he worked with, several of whom urged him to tell this story, but it took cajoling to get him to tell the story. This is an organization in which humility is highly respected and members—perhaps especially Chief Petty Officers—are uncomfortable openly drawing attention to their own personal heroics. Indeed, there is an art to telling epics, drawing attention to oneself as the hero of an epic can backfire in an organization like the Navy that values humility and team work over individual accomplishments—you cannot organize your own ticker tape parade.

Epic stories usually cast a hero, who can be an individual or a group, in the role of the protagonist. Epics also usually include someone or something that needs to be rescued from danger of some kind. Epics describe contests, trials, challenges, quests, or tests. The themes of epic stories are usually courage, nobility, loyalty, selflessness, honor and self-sacrifice. The heroes in the epics are often the storytellers themselves, but sometimes the heroes are others whom the storyteller admires. The heroes in the epics usually surmount the challenges they face through skill, determination, perseverance, courage, sacrifice, wit or ambition. The emotions evoked by the epics are pride, admiration, nostalgia and sometimes envy—admiration was a much more prevalent emotion than envy in the stories of Navy Chiefs.

Senior Chief Tim used *USS Kitty Hawk Mishap* to make sense and cope. He and the organization have used the story of this incident, captured on video, to informally train others. *USS Kitty Hawk Mishap* reflects not only the values of the organization but Senior Chief Tim's values as well. The telling of this story allowed others to vicariously participate in a very scary and dangerous situation from a safe distance. Senior Chief Tim faced a challenge—a test—

saving the injured pilot who was in shock. He tackled the challenge with courage, selflessness and strength. The story evoked admiration in the audience and pride in an organization that cultivates such virtues.

Senior Chief Tim told the story well. He imbued the scene with just enough drama to communicate the story's meaning without seeming boastful. The events described did not need much additional dramatization and Senior Chief Tim seemed to sense this. Senior Chief Tim described how the information received by his senses—the "BAM" and the "orange flash out of the corner of his eye"— did not make sense at the time, but in retrospect he was able to put the actions together in a logical sequence that did make sense. He described a plot—a problem—the aircraft crashed, it was in flames, the pilot was in shock and stuck in the seat pan. And he described how he instinctively reacted to rescue the pilot. Senior Chief Tim made the interesting observation that "some people just freeze when something like that happens."

Stories like *USS Kitty Hawk Mishap* allow people to mentally simulate and contemplate how they might react in such a situation. Even though no two incidents are identical, there are similarities and no doubt the narratives that have formed the cultural fabric of an organization influence how individuals will react when faced with similar challenges. It is not known precisely how specific stories will effect the future actions of the individuals who hear them but they no doubt do.

The majority of organizational stories are epics (Gabriel, 2000). The preponderance of epic stories is to be expected. Life presents all of us with daily challenges, trials and tests that become the plots of personal and organizational epics. Of course, most organizational epics are not as extreme as *USS Kitty Hawk Mishap*—even in organizations that have dangerous missions—but all organizations and individuals have their epics. Epic stories as a genre are a perfect example of stories that are used to make sense of the past and pass on that sense to others. But epics also help people cope, and they function as maps for future behavior.

The story *Page 10's* told by Master Chief Tom in Chapter 1 is a less extreme example of an organization epic than *USS Kitty Hawk Mishap*. In *Page 10's* Master Chief Tom described an incident that was like incidents he or others had experienced or would experience in the future. The story provided behavioral maps for how to react to similar situations. *Page 10's* provided a model of followership—it described a novice willing to work hard and learn at the foot of the master. And it also painted a picture of leadership—a leader willing to share his knowledge and take the time to show a novice how to complete a task.

Most people who have worked in business organizations will be able to relate to the following epic told by Master Chief Arcelo, a logistics specialist who was in charge of getting work done in the Aviation Intermediate Maintenance Department (the department that repairs broken aircraft equipment) on the Ship. This epic is less extreme than *USS Kitty Hawk Mishap*, but Master Chief Arcelo, like Senior Chief Tim, was a hero in this story none-the-less, he arrived just in time to turn a tragic narrative that had become part of the fabric of the organization into an epic achievement.

WE'RE GONNA ACE IT!

When I reported to my last ship morale was really low in AIMD [Aviation Intermediate Maintenance Department]. They had just failed their AMI [Aviation Maintenance Inspection]. I heard all these stories about how bad they were. After the ship won the Battle E [an award for excellence] the Captain came down to the Chiefs' Mess and he asked every department to stand up for a standing ovation with the exception of AIMD because they had failed. They felt really singled out and bummed out. And so when I came in they were preparing to be re-inspected. And everyone was like, "We're gonna fail again." They were a sorry group. They were acting hopeless. And I come in and says, "I have never failed an AMI and I am not gonna start now. We're Gonna Ace It!" And they were like, "Yeah right, how we gonna do that?" And I started to look at things and we started to make some changes and I just kept saying, "We're Gonna Ace It!" And pretty soon they were like "Okay, maybe we can pass." And I was like "We're Gonna Ace It!" And as we got closer to the re-inspection I think they actually started to believe that it was possible. And when it came time for the re-inspection we aced it.

Master Chief Arcelo described an organizational experience that many of us have lived through. He faced a challenge. He rallied the team through skill, determination and perseverance. Like the hero in a classic epic he came to the rescue of the group. When asked how he knew the department was going to "ace it," he chuckled and said, "Hell, I didn't know, there were times when I had my doubts, but I didn't let them know that."

It may seem like common sense, but often organizations—and leaders—do not recognize, much less use the epic struggles that have formed them to craft stories that will help them succeed in the future. Used effectively, epic stories like *We're Gonna Ace It*, *USS Kitty Hawk Mishap* and *Page 10's* can play a hugely beneficial role in organizations. They generate pride in the organization, admiration for the hero and they inspire others, consciously or unconsciously, to emulate successful behavior. They communicate organization values and provide behavioral maps. Epics also have a happy ending, and who does not like a happy ending?

If organizational epics are so powerful where can they be found? Some organizational epics are immediately obvious, and there are some exercises at the end of this Chapter that might help uncover personal and organizational epics that are not immediately obvious. But wise leaders need to be self aware and secure enough to realize that they will not always be the heroes of their organization's epics. Sometimes—like Bilbo Baggins in the *The Hobbit*—heroes of epics are found in unlikely places. The daily news is replete with stories about ordinary individuals who accomplish extraordinary feats of heroics when confronted with a crisis. Senior Chief Tim was a young first class petty officer when he rescued a pilot he didn't even know in *USS Kitty Hawk Mishap* yet he took action while others with more training and more experience momentarily "froze." Master Chief Tom looked inside himself to find an epic story that took place 30 years earlier about a mentor that could be instructive to others today.

The lesson for organizations and leaders is that—like the bird in the apex of a migrating flock—sometimes those in positions of leadership need to step back and allow the epics and their heroes to emerge. Epics reflect many laudable values of an organization and culture including: selflessness, determination, commitment, stamina, team effort, creativity, wit and achievement in the face of odds. Used effectively, epic stories play an important, beneficial role in organizations. They generate pride in the organization, admiration for the hero and they inspire others, consciously or unconsciously, to emulate successful behavior. They communicate organization values and in that way they are instructive.

LEADERSHIP REFLECTIONS ON EPICS

1. *Think of a time when you, or your organization, faced a challenge or a trial of some kind and prevailed. Describe the problem and how it was solved. Articulate a story surrounding that experience with a beginning, middle and a successful conclusion, or lesson learned. What does your story say about your values or the values of the organization? How has that experience influenced how you, or your organization, tackles challenges today? How might you use your story to train or inspire others?*

2. *Ask a subordinate, an employee, a potential employee, a colleague, or a mentor, to tell you about a challenge they faced and how they prevailed. What insights into the individual, and/or the organization, does the story contain? What does their story say about their values? Do their values mesh with the values of the organization? How might their story be instructive or inspirational to you or others?*

3. *Listen to, and note, the epic tales that circulate in your organization—you will hear them in meetings, presentations and casual conversations. What challenges do they describe? Who are heroes in these epics and why are they heroes? What dangers has the organization faced in the past and prevailed? How might the organization use epics from the past to cope with challenges in the present, or in the future?*

4. *Are there unlikely heroes in your organization? How can you help them feel empowered so that they can achieve their full potential? Can you tell their story to inspire others? Can you craft epics illustrating the achievements of the sometimes unsung heroes in the organization? How can such stories be used to make members of the organization feel empowered? How can such stories be used to inspire future behavior?*

5. *Ask members of your organization to tell you about the people who have inspired them or influenced who they are today? How can these stories be used to illustrate values, principles and behavior that will sustain the organization and move it in a positive direction?*

(See Appendix D (Storytelling guidelines) to hone storytelling skills)

Chapter 4

COMEDIES

Congratulations to our new Chiefs!

You gotta remember you were once standing there with nothing but two little French fries on your sleeve.

Master Chief Hassan

Daniel Goleman in his emotional intelligence research found that the artful use of humor typifies effective leadership (1995). Humorous stories are told to relieve pressure or entertain an audience and thus help people cope. They are usually cathartic or energizing in some way. Most of the humorous stories told by Chief Petty Officers were like inside jokes; that is to say they would not be funny—or as funny—to those outside of the organization. Many involved self-deprecating humor. Some of the stories described passive aggressive behavior and some contained grim or ironic themes. Some had a "gallows humor" quality.

The following story, told by Master Chief Tom, the curriculum director at the Senior Enlisted Academy who also told *Page 10's,* is more humorous to insiders than outsiders, but even outsiders can appreciate its humor and its ability to expose the sometimes absurd nature of the pomp and circumstance that surrounds formal traditions in a hierarchical organization that shows great deference to rank and position. In this story Master Chief Tom referred to being a "side boy," that is to say part of a formation of sailors in dress uniform that is posted at the ceremonial quarterdeck of a ship to render honors to dignitaries when they come on board. As a dignitary crosses the quarterdeck the ship's Boatswain Mate blows his pipe to announce their arrival and the side boys snap to attention—in perfect unison—rendering a sharp hand salute. They remain at attention, saluting, with their "eyes locked" (looking straight ahead) until the dignitary has passed between the side boys and has been escorted through the ceremonial area.

DIPPY SIPPY DONUT GUY

I was on the USS Forrestal, and it was my first experience as a side boy in an honor guard. It was a nice day and we were having a Change of Command, which as you know is a huge deal for an aircraft carrier. I was proud to have been chosen to be a side boy and I was lookin' particularly sharp in my cracker jacks [sailor's dress uniform]. We were expecting all these dignitaries— senators, admirals, generals, ambassadors of one sort or another—you know the drill. And the plan was that there would be a lieutenant stationed down the pier with a radio. And when the lieutenant saw a dignitary coming he would radio the quarterdeck so the Bosun [Boatswain Mate] would know someone was approaching so we would be ready. We had done a couple of practices. And of course we were out there early all ready to go. So, all of a sudden the Bosun [Boatswain Mate] blows his pipe, and I was thinking, wow the dignitaries are arriving early. And we snap to attention and render a sharp salute and I've got my eyes locked [looking straight ahead] just like I've been taught to do and out of the corner of my eye I see this guy in a white uniform walking up the brow and he is carrying something big and white and as he passes in front of me I see that it is this pimply faced teenager carrying these big white boxes. And it turns out he is the delivery boy from the Dippy Sippy Donut shop who has been sent over to deliver pastries for the reception. And we had just rendered him full honors as if he was an ambassador or something.

Chiefs enjoy sharing stories like *Dippy Sippy Donut Guy*, and they have many such stories. The telling of one humorous story usually prompts the telling of another as the audience recalls similar experiences. Humorous stories like *Dippy Sippy Donut Guy* are the kind of stories that are usually told in a casual setting, among peers. They are energizing, they help people relax and they serve to increase the bond of the group. Storytellers who can make people laugh are highly esteemed in the Navy as well as most other organizations.

The story *Dippy Sippy Donut Guy* was humorous because it exposed the irony of a formal tradition. The pomp, circumstance, and strict hierarchy that is such a part of Navy traditions was the brunt of the joke. A lot of time and extra effort goes into preparing a ship for traditional ceremonies and such formalities are taken very seriously, especially by the organization's leadership. Sailors usually grumble about the extra effort that goes into preparing to receive "dignitaries" and it is the Chief Petty Officers who must motivate the crew and keep them on task to put in the extra effort. But when all is said and done, the ship sparkles, the crew looks sharp and there is a general sense of pride and accomplishment. Traditions serve a valuable purpose in the Navy. They remove some of the uncertainty from an environment that can quickly become chaotic. They allow people to operate in a synchronized fashion automatically. They reinforce behavior and attitudes that are valued in the organization such as respect and attention to detail. And as long as traditions work they are invisible but when they fail, like slap stick humor, they result in surprise and sometimes hilarity as the organization takes a moment to laugh at itself.

Through *Dippy, Sippy, Donut Guy* Master Chief Tom was connecting with his audience by creating a bit of levity surrounding something they have all endured. The story helped them laugh at themselves and strengthened the bond of the group. It was ironic but amusing that a pimply-faced delivery boy who was probably the same age as the lowliest seaman on the ship would receive the full honors of a dignitary.

Like *Dippy Sippy Donut Guy* many of the humorous stories told were self-deprecating. Daniel Goleman (1995) in his research on emotional intelligence found that effective leaders are often masters of self-deprecating humor. Leaders who are self aware and mature can laugh at their own foibles which helps to foster an atmosphere that communicates that it is okay to take risks and make mistakes—especially if the individual or the organization learns a lesson from the experience. Such stories communicate that an organization that can laugh at itself is an organization that is open to creativity and growth.

In the following story Master Chief Hassan describes arriving at the ship with two little "French fries" on his sleeve—narrow half chevron stripes indicating that he was right out of basic training—a lowly seaman apprentice. He also describes a Navy custom of striking the ship's quarterdeck bell four times in succession whenever the captain of the ship arrives or departs the ship. Anyone who has spent even a short period of time on a Navy ship is familiar with the daily routine of announcements over the "1MC"—the public address system. Such announcements are one of the ways the crew receives important, as well as routine, information. When the Commanding Officer departs the ship the bell is struck four times followed by the announcement that the Commanding Officer of the ship is departing, but the quarter deck watch officer always uses the name of the ship—not the Commanding Officer's name—when making this announcement. Hopefully that explanation will make the following story—that is hilarious to insiders—understandable to those who have not served in the Navy.

FRENCH FRIES

When I reported to my first ship the USS Marvin Shields, I was a lowly E-2 [seaman apprentice]. I walked across the brow and I was standin' there with my big old sea bag and my two little French fries (Hassan indicates two little lines with two fingers on his upper sleeve)—you got the picture [lots of chuckles and head nodding] and they point me on my way to get down to my rack in berthing [the crew's sleeping area] so I can stow my gear. And I go down there and I've got this "coffin rack" [stacked bunks attached to the wall with hardly any headroom] and this little box of a locker to stow all my gear in. And I am trying to fit all my stuff into a locker when I hear over the 1MC [the public address system] "BONG, BONG—BONG, BONG—MARVIN SHIELDS DEPARTING," and I says to myself "Whoa, Marvin Shields departing, I got on board just in time, we're getting underway, that was a close call." And I am waiting to feel the ship rockin' and rollin' or something and I don't feel a thing and I think "this ain't so bad, I don't feel a thing." So I continue getting settled and about an hour later I hear "BONG, BONG—BONG, BONG—MARVIN SHIELDS ARRIVING," and I am thinking "Wow, that was a short trip!" And a couple of other guys come into the compartment [room] and I say to them "Hey, where did we just go?" And they were like, "What do you mean where did we just go?" And I said, "Well they passed the word that Marvin Shields was departing." And they say "You freakin' boot camp idiot…they say that every time the captain leaves the ship." And so I say, "Whoa, that's pretty freakin' cool!" And they say, "What's so freakin' cool?" And I say, "Well, here I am on the USS Marvin Shields and the captain's name is Marvin Shields—that's a freakin' amazing coincidence."

The group burst out laughing when Master Chief Hassan told the punch line to this story. It was funny enough that an articulate, accomplished Master Chief did not understand a protocol that is a ubiquitous part of shipboard life when he was a naïve seaman apprentice—there were others in the group who obviously could relate to that. But when he finished with the punch line— like all good comedians—he took everyone by surprise as they too saw the "freakin' coincidence" through the eyes of a young seaman apprentice. Master Chief Hassan was holding a mirror up for all in the group to see themselves as they once were and as they are today. He was allowing the group to see the world—their world—through the eyes of a young seaman apprentice once again and there is tremendous value in that. How can a leader, a mentor or a supervisor communicate with 19 to 20 year olds if they cannot remember and relate to what it is like to be that young and naive? It is the wise and mature leader, who has made sense of his past, who can admit that he was once "a boot camp idiot." After telling the story Master Chief Hassan had the rapt attention of the group every time he spoke.

Many of the stories that the Chief Petty Officers told had a "gallows humor" quality—that is to say they described grim or ironic situations that were probably not funny at the time but in retrospect were humorous. Gallows humor provides members of the organization with "protective armor" (Gabriel, 2000, p. 88) that helps them cope with the physical and emotional demands of the job "such stories often reveal a proud and defiant protagonist who rejects self pity, making light of the hardships turning victimhood into survival against the odds if not outright victory" (Gabriel, 2000: 88).

In the following story Master Sergeant Martin, a marine NCO (Non Commissioned Officer) who was an exchange student at the Senior Enlisted Academy, used a bit of gallows humor to make a point. He described an incident that took place while "one of his buddies" was on sentry duty at military base in California. It is the job of the gate sentries to check the identification of vehicles and individuals coming on to the base to ensure that all who enter have proper authorization. The cars of Commissioned Officers display a special windshield sticker that indicates that the owner of the vehicle, who is most often the driver, is a "superior officer." Enlisted gate sentries render a sharp hand salute to all of the officers entering the base while they simply "wave on" the cars of enlisted personnel entering the base.

GOOD MORNING STICKER

This buddy of mine was stationed at El Toro and he was working the gate [granting or denying access to the base]. It was a typically busy morning with lots of traffic and a long line of cars coming onto the base. So anyway, this officer's car came up to the gate and my buddy notices that there was a woman in civilian clothes driving, who just happened to be a colonel's wife. Anyway he was in hurry, it was rush hour, the line to get on the base was getting longer and he waved the car through the gate. When he waved the vehicle through the woman driving the car slammed on the brakes. And she rolls down her window, points at the sticker and snaps at the sentry, "Didn't you see that sticker?" and he says "Yes, Ma'am," and she says "Well, you are supposed to salute that sticker, aren't you?" and he says "Yes, Ma'am!" And he bent over looking right at the sticker, gave a sharp salute, and said "GOOD MORNING STICKER! HAVE A NICE DAY STICKER!"

This story was rich. It no doubt has been used before to entertain, help members of the organization cope and even prescribe a map of how to behave, or cope, in the future. Everyone in the group chuckled when Master Sergeant Martin told the punch line. They all had experienced, or heard stories before about, officers' spouses who "wore their husbands' rank on their knickers"—a phrase often used to describe imperious wives of superior officers who flaunt their husband's rank. This was a passive aggressive story. It described an ironic situation—the requirement to salute a sticker in an organization that is highly rank conscious. It is interesting that Master Sergeant Martin did not caste himself in the protagonist's role in this story. The protagonist was a "buddy." My hunch is that this story is well traveled among gate sentries in the Marine Corps. And it is an entertaining story to tell those outside of the organization as well because who does not like to see a silly, imperious person get their comeuppance? Perhaps the story was an accurate reflection of exactly what the gate sentry did, or perhaps the story reflects a fantasy, that Master Sergeant Martin and numerous other Marine gate sentries wish they could have, or would have, done under similar circumstances. Or perhaps Master Sergeant

Martin was reluctant to tell the author (a retired, female, Commander) that it was actually he who experienced the incident. Either way, the story is a rich reflection of a culture and once again a member of the group was holding up a mirror so that the organization could see how silly it is at times.

But humor can be tricky—one person's joke can be another person's insult. While humorous stories can relieve tension, energize a group and be instructive, they can also be demeaning and mask aggression or scorn. Stories that elevate the status of one group, or one individual, at the expense of another can debilitate not only the individual or group that is the brunt of the joke but an entire organization. Humor used to demean an individual or a group of individuals at the expense of another group can create harmful tension, sap energy and debilitate an organization. That being said, self-deprecating humor is the mark of mature and confident leaders who are not afraid to laugh at their own foibles, or the foibles of the organization

LEADERSHIP REFLECTIONS ON COMEDIES

1. *Describe a mistake that you or your organization made that in retrospect is really rather humorous? What does this story say about the organization?*

2. *What humorous stories are told and retold in your organization? How do these stories reflect the values of the organization? Do these stories create bonds in the organization or do they marginalize individuals or groups of people?*

3. *Who are the people in your organization who have the ability to make people laugh? What does their brand of humor say about the culture of the organization? Where are the places in your organization where you hear laughter?*

4. *Think about your mentors, or leaders you admire, how did they use a sense of humor to make people feel comfortable or motivate and energize them?*

(See Appendix D (Storytelling guidelines) to hone storytelling skills)

Chapter 5

TRAGEDIES

Bad news travels fast and it's never accurate.

Senior Chief Evers

The tragic stories that Chief Petty Officers tell are usually about accidents—called "mishaps" in the Navy. Organizational tragedies are usually about the misfortunes of others. Unlike epics or comedies, the storytellers rarely cast themselves as protagonists in tragic stories. Often the protagonists in tragic stories are non-deserving victims suffering from some sort of misfortune, insult or injury. Tragic stories do not have happy endings, but that is not to say that they do not have positive outcomes; there are usually valuable lessons or morals contained in seemingly tragic stories for those open to such interpretations.

Tragic stories circulate through most organizations. They are often told as cautionary tales. They may be told in whispers and spread with lightning speed through the water cooler web—called the "scuttlebutt" in the Navy—or they may be widely disseminated through official channels as examples of what not to do. They may be about failures within, or outside of, the organization. Some tragic stories are terrifying, allowing the audience to rehearse their reaction to a horror at a safe distance. Some tragic stories identify scapegoats that provide the organization with an explanation for the tragedy, thereby enabling the organization to cope and move beyond the tragedy.

In the following story, Senior Chief Tim, the Aviation Boatswain Mate who told *USS Kitty Hawk Mishap*, communicated the importance of the work that the Aviation Boatswain Mates do to a group of young sailors on duty in the flight deck control compartment. The Aviation Boatswain Mates who work on the flight deck are out on deck in all types of conditions; extreme heat and cold, rain and snow. They work in twelve-hour shifts. The work is physically demanding and dangerous. One wrong move while they are directing a plane can result in physical injury, loss of life, and the loss of multi-million dollar equipment. In the following tragic story, Senior Chief Tim refers to "stroking"— the term used by Aviation Boatswain Mates to describe the hand and arm movements used to direct the aircraft around the deck. "Stroking" movements are standardized, international, signals. It is important to understand that although pilots have the high profile job of flying the planes when they are in the air, once they are on the deck they are guided by the Aviation Boatswain Mates assigned to direct the planes off the flight line and into a position where they can be moved below to the Hangar Bay. The Aviation Boatswain Mates who direct the aircraft are called "handlers."

STROKING ON HIS DEATHBED

Jack and I were really good buddies. We grew up on the flight deck together. We had some awesome times. Anyway, for one reason or another we both left the Navy after our first enlistments. Jack got married. He and his wife were both from Iowa and he wanted to settle down there and start a family—the whole nine yards. His wife, Tiffany, was the real deal. Anyway, I ended up coming back into the Navy. I missed it. We kept in touch even though I was bouncing around the world. Sometimes I wouldn't talk to Jack for months at a time but then I would. Anyway, he got leukemia, and I didn't even know it. And one day I got a call from Tiffany and she told me Jack died. And she said she thought that

I would want to know that he was talking to me on his deathbed. She said at times he was delirious and he kept saying my name and talking to me and he was "stroking." She said she thought I would want to know.

Stroking on his Deathbed is a tragic story but it is not about a personal tragedy that the storyteller experienced himself. Although Senior Chief Tim lost a dear friend he was not the victim in this story. Through telling this story Senior Chief Tim was making sense, and giving sense, to those who were listening to the story. He was explaining the decision he made to return to the Navy. The implication was that Jack's work in the Navy was such an important part of his life, and his connections to it were so strong, that in his dying moments he was back on the flight deck "stroking." Through this story Senior Chief Tim was telling the young sailors present that they are in a special profession that they should be proud of. He was communicating to the sailors listening to the story that they are part of something bigger than themselves. Senior Chief Tim was giving the sailors who heard the story something to sustain them on the days that they are working so hard that they are ready to drop from exhaustion, or the days that they are bored stiff because there are no planes flying. He was communicating that they should be willing to put up with the hardship that comes with the job because they are part of a noble profession—they are the "handlers." He was telling them that the friendships and bonds that they have with their shipmates are stronger and will endure longer than most friendships. He was communicating that they should feel proud and privileged to be part of a special group.

In the following tragic story, Senior Chief Evers, who was in charge of the division that maintains the arresting gear—the equipment that catches the jets when they land on the deck of the ship—describes the failure of a piece of equipment that resulted in the death of one sailor and the maiming of another.

JET BLAST DEFLECTORS MALFUNCTION

We had been having trouble with the Jet Blast Deflectors. We had technicians onboard in the shipyard repairing them and when they reconnected the hydraulics lines to the Jet Blast Deflectors they did it backwards. Two of my guys were trying to figure out what was wrong with them and in the process of trouble shooting, the equipment malfunctioned and they both ended up pinned (crushed underneath the heavy equipment). One of the guys was killed instantly and the other one was stuck and he was calling out to me by name. I ran over to him but I was powerless to do anything. It was a sick feeling—this guy calling my name and I couldn't do a thing to help him. We eventually got him out but he lost an arm and a leg. And after spending all day in the emergency room, I come home, walk through the door, and my wife who was pregnant at the time sees me, screams, and collapses on the floor. Bad news travels fast and it's never accurate. She had heard about the incident, heard there were fatal injuries, knew it was my division and thought it might be me. I should have called her to tell her I was okay, but that was before everyone carried around cell phones. She was so stressed out she went into pre-mature labor. So there I was back in the emergency room again. It was the worst day of my life. But I guess when all was said and done some good did come out of it—as a result of that incident the Jet Blast Deflectors were modified on all aircraft carriers so that there was no way you could reverse the hydraulic connections, so an incident like that will never happen again—but I lost two of my best guys and my wife almost lost the baby. Like I said, It was the worst day of my life.

Tragic organizational stories are not as common as epics or comedies.

Storytellers are usually uncomfortable sharing the intimate emotions that such stories evoke, such as sadness, love or fear—especially with a stranger. Personal tragedies are the kind of stories people may tell to intimate friends, family, or perhaps mental health professionals, but often they keep their personal tragedies to themselves.

Organizational tragedies like the two used as examples here can be instructive to others. And the Navy does use cautionary tales based on mishap investigations in training. The lessons of *Grampaw Pettibone*, a fictional naval aviator who tells stories about aviation mishaps, or near mishaps, are standard fare in Navy Safety publications. But a story told in an informal setting by someone who was actually there is often more memorable. It is likely that an Aviation Boatswain Mate who has heard *Jet Blast Deflectors Malfunction* will recall the story when working with or around the Jet Blast Deflectors. This story may cause someone, consciously or unconsciously, to be more cautious. That being said, there are some organizational tragedies that are simply too intimately painful to share with outsiders, or even non-intimate insiders, and they are best left to a counseling session with a trained mental health professional, or a trusted friend, who can supply the emotional support needed to help the protagonist make sense and move on.

There is a type of personal tragedy story that parents sometimes use quite effectively to instruct their children. It is the type of story in which a parent describes a stupid mistake and explains to the child that they are sharing the story out of love because they are concerned that the child might make a similar mistake and get hurt. Parents must be mature and insightful to openly admit to their children that they are fallible, but when they do they do they inevitably capture the attention of their children. Mature and insightful leaders may also effectively employ such stories. A leader who feels comfortable sharing a story about a personal mistake, or failure, not only shares a cautionary tale that can help others, but also communicates genuine concern for the welfare of the members of the group. The intimate bond that such stories create fosters cooperation and teamwork and can sustain the group through hard times.

Master Chief Eric, a submariner, calls cautionary tales that mimic past tragedies "Don't be the guy" stories as in "Don't be the guy who (fill in the blank)..." Master Chief Eric's explanation follows,

DON'T BE THE GUY

On my last boat [submarine] we had this kind of story that we told called "Don't be the guy..." stories. We used them when we were preparing the crew for one evolution or another. We used "Don't be the guy..." stories to sort of anticipate things that could possibly go wrong and a lot of them were based on personal experiences. I remember one time when we had a bunch of midshipmen coming on board from the Naval Academy and some of them were women. And you know, we got a young crew, and submarines are small and cramped [compared with other warships], and it's still a male dominated environment, so having a bunch of young females from the Naval Academy on board is tricky. So the COB [Chief of the Boat] got the Chiefs together and we talked about how we were going to deal with the situation and we came up with all these "Don't be the guy who..." stories to tell the crew to hopefully prevent problems. It's like "Don't even think about it," or "Don't even go there," but of course just saying that makes you think about it and think about the consequences which can be a good thing. Sometimes a little bit of fear ain't so bad.

Depending upon their interpretation, organizational "tragedies" can be devastating events that the organization may never fully recover from, or they can be rich cautionary tales that strengthen the emotional bond within the group and help the organization move in a positive direction. Unfortunately, there is a tendency within organizations to gloss over, cover up or ignore tragedies. Tragedies are painful, it is a natural both personally and within organizations to shrink from pain but ignoring tragedies once they have occurred is counterproductive. Crafting an authentic narrative that makes sense of tragedy can be cathartic, allowing individuals and the group to heal, learn and become stronger. Ignoring organizational tragedies, whether real or perceived, will not make them go away and may increase their debilitating effects. With the benefit of hindsight most tragedies can be interpreted as epics or even comedies. But some tragedies will remain tragic and their value is in the lessons they contain. A wise leader, through narrative, makes sense of a tragedy, which allows the organization to cope, learn, grow stronger and move forward.

Also, a wise leader who is in tune with the narratives coursing through the organization can sense when members suffering from personal tragedies need professional help. A cultural stigma persists in the military, as well as in corporate America, against referring members of an organization to mental health professionals. The narrative that persists in many organizations is that the need for professional counseling is a sign of weakness and even the "kiss of death" in an otherwise promising career. But just as individuals suffering from physical disabilities can be healed, members of an organization suffering from mental and emotional disabilities can be healed—their personal tragedies can be turned into epics. Mental health professionals can help members suffering from mental and emotional disabilities deconstruct the complex narratives that are making them ill and construct coping narratives that will help them move in a positive direction. When one member of an organization is suffering the entire organization will suffer to some degree. When members who need help receive assistance they can often be returned to productive service and the entire organization benefits in ways that are discussed in greater detail later in this book.

LEADERSHIP REFLECTIONS ON TRAGEDIES

1. *Recall an organizational "tragedy"that you, or the organization, experienced, what lessons were learned? Was it really a "tragedy" or in retrospect could it be interpreted as an "epic" or even a "comedy"?*

2. *Are there perceived "tragedies" within the organization that are inhibiting risk taking, creativity and growth? How can you as a leader make sense and be wiser as a result of these tragedies and communicate that wisdom to others?*

3. *Are there mistakes that you, or your organization, have made that you need to recognize and make sense of? What can you or the organization learn from stories about mistakes? What can you, or the organization, learn from the tragedies that have taken place in other organizations?*

4. *There are people in all organizations who have experienced, or will experience, personal tragedies. Are you as a leader sensitive to the needs of individuals who may be suffering from personal tragedies in silence? Does your organization have access to mental health professionals who can help members deal with personal tragedies and move in a positive direction if such help is needed?*

5. *What are the narratives in your organization that are used to prescribe behavioral maps when tragedies occur? Is the need for professional mental health counseling viewed as a weakness and a career "kiss of death," or is such counseling accepted as a way to move in a positive direction and become a fully productive member of the organization once again. What stories support the prevailing attitudes toward dealing with tragedies in your organization?*

(See Appendix D (Storytelling guidelines) to hone storytelling skills)

Chapter 6

ROMANCES

Although I only worked for him for a short time and that was almost 30 years ago, to this day when I am faced with a problem I often ask myself 'What would Larry Bell do?

Master Chief Tom

Most of the romantic stories collected from Chief Petty Officers were nostalgic or they expressed admiration for others, such as mentors. Some of the stories expressed a desire, on the part of the protagonist or the storyteller, to take care of others such as subordinates, family members, or medical patients. Romantic stories expressed emotions such as kindness, caring, generosity and love. While many of the stories that Chiefs shared had romantic qualities, comparatively few were actually about romantic love affairs between two people, and the stories that were about love affairs often involved sexual escapades that were more comedic than romantic, or they were tragic.

The general perception seemed to be that romantic love in the workplace was unacceptable and therefore it was inappropriate to share stories about romantic love in a workplace setting with an outsider—especially an outsider doing research. But stories about romantic love and lust are highly entertaining for both men and women so my hunch is that they are shared with great regularity to relieve the boredom of the workplace routine. Romantic tragedies function as cautionary tales. Romantic epics may be inspirational. And romantic comedies are entertaining, serving to energize the group.

Female Chief Petty Officers, per person, told more romantic stories than male Chief Petty Officers, which was also to be expected. Women in Western culture have an easier time than men expressing the romantic emotions of kindness, fondness, gratitude and love. Most of the romantic stories told by both male and female Chiefs were about people they admired or they were about people the storyteller had taken care of in some way. The female Chiefs seemed more comfortable with the emotions evoked by romantic epics or even romantic tragedies while the men tended to tell romantic comedies. Several of the male Chiefs told amusing stories with sexual sub-texts about sailors in love, or "in lust", who engaged in outrageous behavior for the objects of their affection. In such stories the protagonists were generally portrayed as lovesick fools but sometimes the protagonists emerged as heroes.

Almost all of the Chief Petty Officers told at least one romantic, organizational story about a mentor whom they admired, someone who believed in them and took care of them—such as *Page 10's*. In male dominated organizations that disdain any hint of "touchy feely" emotions, this type of story is safe and inspirational. Romantic stories make human organizations satisfying places to work. They create a bond within the group and allow for the trust that is necessary if the individuals in the group are expected to collaborate for the good of the organization.

In the following romantic-tragedy Chief Sara, a Hospital Corpsman stationed on the Ship, described an incident that occurred when she was stationed on Diego Garcia, a remote island in the middle of the Indian Ocean that is dominated by a U.S. and British military base.

LORELEI GOES HOME

I was a very junior Hospital Corpsman [medic] at the time—right out of A school [Hospital Corpsman training]. Diego Garcia was my first duty station. There was this girl who worked on the base who died. I was new to the island. I didn't know her. She died of natural causes—it wasn't a homicide, or anything like that. I was assigned to assist the doctor with the autopsy. It was the first time I had ever done anything like that. I had never really touched a dead body before. We didn't have a large medical staff there. And the girl, her name was Lorelei, she was Filipino, and her family was in the Philippines and they wanted her body sent back to the Philippines. It took a while to arrange for the airlift to transport Lorelei's body back to the Philippines, so I was assigned to take care of her body until it could be sent back. Everyday I would check in on her. The doctor who did the autopsy and I would talk about her. We talked about the fact that she was someone's daughter and probably someone's girlfriend. And I felt like I got to know her. I took care of her for about a week. The day she left I combed her hair and got her dressed and I tried to make her look as nice as I could. I wanted her to look like the pretty girl she had been when she was alive, when she got home. And when it came time to take her out to the plane that would transport her to the Philippines, the doctor and I walked out to the plane and watched her get loaded on. And as they closed the door of the plane the doctor said, "We're sending you home Lorelei," and we both just broke down.

The Chiefs whose stories are presented here used story to make sense and give sense. They used story to cope and they used story to prescribe how to behave in the future. As a seasoned Hospital Corpsman, Chief Sara has had to cope with the need to take care of other dead bodies since her experience with Lorelei, and her experience with Lorelei has influenced how she copes in those situations and how she instructs the young corpsmen she works with. In the discussion that followed the telling of this story she went on to say, "You need to always remember that you are dealing with people—even though they are dead—who are someone's son, or daughter, or spouse, or boyfriend. You need to be respectful of that." Most of the romantic stories the Chief Petty Officers told reflected concern for people, respect for people and even love for people that make the organization a humanly satisfying place to work.

The following story is perhaps more epic than romantic but it is included here as an example of a romantic story because it is the classic tale of a sailor willing to go to great lengths for love—or perhaps lust. Its audience found it entertaining. It helped to pass the time on a day when planes were not flying and work was slower than usual, but it is also a rich reflection of the culture and loaded with meaning. This is the type of romantic story that men favored. It is perhaps inspirational to think that there are women worth risking all for; women who could launch a thousand ships. The story was told by Senior Chief Tim, the self described "fully posable GI Joe with kung fu moves" who looked fully capable of pulling a Johnny Weismuller stunt like the young, protagonist in this story.

JOHNNY WEISMULLER IN SUBIC BAY

We were in Subic Bay, Philippines. You've been there so you know what I am talkin' about. You know sailors love Subic Bay. So we had been there a couple of days and we were getting ready to get underway. And you know you always leave a couple of guys in Subic Bay. Anyway we were getting ready to pull away from the pier and we can see this guy by the gate in a lip lock with this girl. And we are shouting at him to come on, but this guy is not coming up for air. We were shouting at him and the ship started to pull away from the pier and finally he breaks away from the girl and starts running. By this time we were a good ways away from the pier. And this guy comes running down the pier and does this like Olympic dive off the pier and starts swimming out to the ship like a bat out of hell. And we are all cheering him on. And he is swimmin' like Johnny Weismuller—ya know in the old Tarzan movies—he was smokin'. By this time quite a crowd has gathered on the ship and on the pier and we are all cheering him on. And he actually catches up with the ship—which was amazing. And he somehow manages to climb on board. And we were all cheering and clapping that he made it. It was amazing. But of course he went to Captain's Mast [disciplinary proceedings] for missing ship's movement. I think the skipper should have at least given the guy "A" for effort. I don't know, maybe he did. Anyway I was talking to him one day a few weeks later when he was chippin' paint, paying for his love life, and he said, pardon my French, "Fuck it, she was worth it."

Johnny Weismuller in Subic Bay, is a classic "sea story". Sailors have been jumping off ships and missing ship's movement for love, or lust, at least since the days of Odysseus. Such stories are legend. Romantic epics about girls who were "worth it" fuel fertile imaginations and add spice to life. Most are benign and even energizing; they no doubt help pass the time at sea when nothing much is happening. But there were also romantic tragedies about sailors who attempted suicide when they received "Dear John" letters, or foolishly risked their lives jumping ship in a vain attempt to re-unite with the object of their affection. And, of course the stories about sailors who have gotten into bar fights over women are legend. Whether epic, tragic or comic, stories about romance are entertaining, sometimes titillating, and even inspirational, although sometimes they inspire behavior that is not in the best interest of the organization and sometimes they are nothing more than salacious gossip that can be hurtful to both individuals and the organization. When stories about love, or lust, reflect gossip or behavior that is, or could be, damaging to the individual or the organization, the wise leader has the courage to intervene in a timely manner and expose them for what they are.

While the workplace may not the appropriate venue for a romantic love affair, it is normal for humans to develop bonds with the people they work with. Genuine fondness and respect for other humans make organizations satisfying and safe places to work. Stories about love and lust will be told in human organizations; some are benign entertainment and some are even energizing, but others can be energy drains and damaging to the individual or the organization. Just as some comedies mask aggression or scorn, some romantic stories mask jealousy and even hatred. That being said, they can, and will continue to be told, in human organizations and are rich reflections of organizational values and culture.

LEADERSHIP REFLECTIONS ON ROMANTIC STORIES

1. *Describe specific incidents that reflect the genuine concern for, and respect for, the people in your organization. What insights do these stories contain about the organization and the people in it?*

2. *Remember that everyone is someone's loved one, take the time to tell a subordinate or colleague's family, how much you appreciate their loved one— describe something specific they did to win your admiration. Remember to tell your own loved ones how much you appreciate their love and support— cite specific examples.*

3. *Tell a story about one of your mentors that illustrates why you admire them, think about how you display the admirable qualities you learned from them. How might the stories about your mentors be instructive to others?*

(See Appendix D (Storytelling guidelines) to hone storytelling skills)

SUMMARY OF PART II

Humans are constantly engaged in a process of organizing and interpreting their thoughts, knowledge and information in an attempt to make sense and cope. Humans interpret what is happening in and around them through story and narrative. Story and narrative are such a ubiquitous way of making sense that—like breathing—humans rarely recognize "storying" when they are doing it. It takes a conscious effort to recognize stories, and the larger organizational narratives that they inform, to tap into the power and potential of story and narrative. One way to begin to identify the stories and narratives that have formed the characters of individuals and the culture of organizations is to provide a structure for the interpreting them. Patterns and themes begin to emerge when personal and organizational stories are classified into genres such as epics, tragedies, comedies and romances and hybrid combinations of those genres such as epic-comedies or tragic-comedies. Once stories are evoked by examining genres it is easier to see how they help the individuals who tell them and hear them make sense, cope and navigate into the future.

The majority of organizational stories—such as *USS Kitty Hawk Mishap* and *We're Gonna Ace It*—are epics. Gabriel (2000) explained that the bulk of organizational stories fall into the epic mode because epic stories are entertaining and safe to tell an outsider. Gabriel said,

The emotional content of epic stories is quite distinct from those of comic and tragic. They invariably generate pride and admiration. They also generate commitment and even a sense of duty to emulate the hero or maintain the tradition (2000: 74).

Stories with strongly tragic elements usually are not about the storyteller's personal tragedies, they are stories about the misfortunes of others or accidents—called mishaps in the Navy. *Jet Blast Deflectors Malfunction* and *Stroking on his deathbed* are examples of tragic stories that reflect the misfortunes of others. Storytellers are often uncomfortable sharing stories about personal tragedies with someone they do not know well because of the intimate emotions that such stories evoke such as sadness, sorrow, pity, fear or anger (Gabriel, 2000). But often tragic stories can be reinterpreted as epics, or as cautionary tales that are cathartic and will allow the organization to successfully move beyond the tragedy. Members of an organization who are suffering from personal tragedies may require the intervention of a mental health professional to cope and move in a positive direction, but the entire organization benefits when the member is restored to health and returned to full productivity.

Stories with comic elements are often cathartic and serve to energize an audience or a group. Some comic stories increase tension while others allow the audience to relax. Many of the humorous stories Chief Petty Officers tell are self-deprecating, self-effacing or they have a "gallows humor" quality; that is to say they describe grim or ironic situations that probably were not funny at the time but in retrospect are humorous. Wise leaders and healthy organizations can laugh at themselves.

Many of the stories collected had romantic qualities, but very few were about romantic love between two people. Most people are reluctant to share stories with an outsider in a professional setting about intimate emotions such as love. Most of the organizational stories classified as romantic expressed deep admiration for others, such as Master Chief Tom's admiration for his mentor in *Page 10's*. Or they were stories that expressed a desire to take care of others such as subordinates, family members, or medical patients. Most organizational stories that involved love affairs or sexual escapades were actually comedies or tragedies rather than romances.

In conclusion, whether epic, tragic, comic or romantic, organizational stories can be instructive and yield insight into organizations and their members. Reflecting upon organization stories using genres to evoke stories and structure the interpretation of those stories is one way to begin to use stories to make sense, cope and successfully navigate into the future.

PART III

LEADERSHIP NARRATIVES

OVERVIEW

Classifying stories by genre—epics, tragedies, comedies and romances—is one way to begin to evoke, interpret and understand the meaning of individual and organizational stories. And, as is evidenced by the discussion of the stories in the previous section, themes, patterns and understandings begin to emerge. But classifying stories by genre, while instructive, does not adequately explain the dynamics of how and why stories work the way they do in organizations. And while insights uncovered by framing stories by genre are worthwhile, they are not profoundly new. A deeper understanding of how and why stories function the way they do becomes apparent when insights from the way natural systems work are applied to an examination of narratives and story—that is what will be done in this section.

INSIGHTS FROM NATURE

Who has not wondered in awe at a flock of birds moving in perfect synchronization around an obstacle, sometimes splitting into two groups to fly or around an impediment, coming together in perfect unison on the other side? Anyone who has drilled as part of a marching unit cannot help but be impressed by the fluid precision of schools of fish and flocks of birds. One might wonder if the lead bird in a formation of a migrating flock gets tired? How does the bird at the apex of the formation pass the word to the birds bringing up the rear that it is time to stop for the night without having all of them plow into one another?

In recent years computer tracking and modeling programs have resulted in insights into how natural systems, including flocks of birds and schools of fish, self organize. It turns out, surprising though it may seem to those who have lived and worked in command and control organizations, the lead bird is not the one who always calls the shots. In fact, the lead bird is not always the lead bird. Organization in natural systems, from cells in a fern plant to schools of fish, emerges without a hierarchy, or a step-by-step plan handed down from on high.

An example of how this works comes from the work of Craig Reynolds. Reynolds was fascinated by how a squawking field of blackbirds near his home would rise in perfect unison and then move off en mass, so he created a computer program called "boids" to simulate the movement of a flock of birds. After months of fine tuning he found that each of the birds in the flock that he created operated on some basic principles, basically: 1) don't get too close or too far from the guy next to you, 2) try to match your neighbor's speed, 3) don't run into anything, and 4) if you are the lead bird in the formation, drop back when you get tired and let someone else take the lead (Goerner, 1999).

The insight from simulations like "boids" is that much of the organization in natural systems emerges spontaneously if allowed to do so, and often there are rather simple principles at work, albeit in a complex world that is constantly changing in large and small ways. Of course instinct—the way non human animals read the environment—plays a major role in animal migration patterns and many of their other adaptations, but the work of Reynolds, and others, supports the premise that living systems are naturally inclined to self organize.

Could this natural inclination to self organize explain why rank and file workers often cannot articulate their organization's mission statement yet they manage to collaboratively get work done? Could this explain why the mission gets accomplished when nothing goes according to plan? Could this explain how an organization, even a country, manages to function after a CEO, or president, or commanding officer, is removed, or replaced? If natural systems, including groups of people, are naturally inclined to self organize what should the role of leaders and managers be—does it mean they are they out of a job? Or are there leadership and management lessons to be learned from the emergence of organization in natural systems? And what does any of this have to do with story and narrative?

Like other biological systems, human social systems can be viewed as entities that have the ability to self organize, and thus maintain themselves. This new paradigm is in contrast to the mechanistic model often used in the past to explain how organizations work. Simply put, the mechanistic model—sometimes described as "reductionist" or "linear"—popularized by Frederick Taylor in the early part of the 20th century fails to adequately explain how human social systems truly work. Applying understanding derived from studying how other living systems work can lead to new insights regarding how human organizations work.

"Storied Spaces" (Baskin, 2008)

We are human beings only in language. Because we have language there is no limit to what we can describe, imagine and relate. Language arose as a result of cooperation. Language gives us identity (Maturana & Varela).

What do insights from biological systems have to do with story? Stories are powerful because they have an adaptable, living quality that cannot be adequately explained in mechanical or linear terms. Stories provide the opportunity for individuals and groups of individuals to interact and play with ideas. The ability to use language to make connections enables humans to be creative, experiment, learn and adapt to changing circumstances and pass on what has been learned to others so that they too can thrive.

Ken Baskin, a writer, consultant and lecturer, who has written about how to apply organic design paradigms to an understanding of how organizations work, applied principles of Complexity Science—a theory that draws parallels between how living systems work and how human social systems work—to narrative interpretation to explain how stories help people discover actions that they must take in order to thrive. Baskin suggests that the idea of "storied spaces… a space defined by the stories we have accepted to explain events that have happened and continue to happen," can be substituted for the concept of complex adaptive systems in nature (Baskin, 2008: 1). He said,

One can think of human social life as an intricate nested network of spaces—family and work group, organization and community, profession and nation—in which membership depends on the acceptance of negotiated stories by which each grouping defines the nature of the world and how people in the group must respond to prosper (Baskin, 2008: 1).

(Note: an overview of Complexity Science theory—a theory that has drawn parallels between human social systems and natural or biological systems—is included at Appendix: A.)

Baskin (2008), and David Boje, one of his colleagues and close collaborators on this theory, have suggested that people use the unfolding narrative—the discourse surrounding a story in the making—as they attempt to cope with a steady stream of new information, communication and stimuli from a variety of sources. This theory supports the notion that the ability to tell stories has been key to the successful evolution of humans—a unique ability that differentiates humans from other living entities. Baskin (2008) and Boje (2001), suggest that story is the way humans organize their thoughts, and that, as individuals, we unconsciously start forming our stories even before we voice them. They suggest that humans articulate their stories and share them with others as they negotiate a personal and collective view of reality. Through the interplay of story, narrative and unfolding—emerging—narratives humans negotiate meaning with others. What is revealed are what Baskin (2008) calls "dominant narratives"—narratives that reflect a collective view of reality that comes to be accepted by an individual or a group as the truth.

Dominant narratives are then used to assess new information; that is to say, they give the individual, as well as the group, a way of responding to the environment. Some dominant narratives appear to be particularly helpful, or resilient, so they prevail for long periods of time with little change; but others disappear, or change, sometimes re-emerging at a later date. Dominant narratives may be replaced or modified by new, emerging narratives. Baskin (2008) found evidence of dominant narratives that functioned like self-reinforcing feedback loops within organizations. He explained that "storying" could allow for the trial and error that is so necessary in the creative and successful adaptation of an entity to its internal and external environments (Baskin, 2008).

The "storied space" (Baskin, 2008) of the Chiefs' Mess

The Navy Chiefs whose stories are explored here occupy a "storied space," that is to say that as a group they share similar values and principles that have come to define their views of reality—who they are and how they respond to the environment that surrounds them. The storied space of the Chiefs' Mess is part of the storied space of the Navy, the military and the country.

As individuals, Chiefs occupy other storied spaces such as their families, their church communities and their regions of the country. Since each Chief occupies different combinations of storied spaces they have unique identities and respond in a unique fashion to their world, but they share the commonality of the storied space of the Chiefs' Mess that largely influences their professional identity—indeed, their professional DNA. The values and principles of the storied space of the Chiefs' Mess are reflected in the individual stories of the Chiefs.

When Navy Chiefs as individuals, and the Chiefs' Mess as a whole, are viewed as "complex adaptive systems"—an entity with the ability to maintain itself—"leadership narratives" that are part of broader dominant narratives within the organization begin to emerge from their stories and the discourse surrounding those stories. The leadership narratives that emerge have attendant prescribed behaviors called "leadership practices" here, which serve to guide the actions of Navy Chiefs and those around them. There was evidence in the Chiefs' anecdotes and their storied interpretations of dominant narratives that are used to define and test reality and thus help individuals, as well as the group, adapt to change and thrive.

How are storied spaces different from organization culture and how are dominant narratives different from cultural values and norms? The ideas of "storied spaces," "dominant narratives," and "emerging narratives" capture the living and adaptable quality of story and narrative that cannot be adequately explained by culture alone. These ideas reflect the uniquely human ability to employ and experience story and narrative as a dynamic process that transcends the examination of anecdote as cultural artifact. Story and narrative can reflect organizational culture as it is today, but they can also reflect and influence what organizational culture might become. An artifact cannot influence what a culture might become until it is imbued with a narrative interpretation that makes it come alive.

There were many "dominant narratives" evident in the stories of Navy Chiefs but there were four dominant narratives that stood out through repetition, and it became apparent that those four narratives largely influence the way Navy Chiefs lead and manage. Because they reflected how Navy Chiefs lead and manage, the four dominant narratives will henceforth be called "leadership narratives." Actual phrases used by Navy Chiefs are used as aphorisms here to capture the essence of these narratives. The voices of Navy Chiefs are used in the following chapters to demonstrate the integral and dynamic role that narrative and story plays in organizations. Most of the stories collected contained more than one of the four leadership narratives and practices, and some contained only one; but the following four leadership narratives and their attendant practices were clearly evident through repetition in the storied space of the Chiefs' Mess.

Through the discussions that came before and after the anecdotes told, the Navy Chiefs illustrated the dynamic qualities of story by connecting the stories to threads in the conversations that came before, pointing out how the story explained something that needed explaining and venturing interpretations of stories that included possible morals or lessons. In this way a wider narrative was emerging that connected stories and anecdotes with the discourse that came before and after them. The emerging narratives were sometimes fragmented, they were the pre-narratives that take place before a complete story is formed (Boje, 2001). A broader narrative unfolded as the storyteller, and others in the group, engaged in making sense of something that did not make sense initially. Members of a group engaged in an unfolding, or emerging, narrative are in search of a sequence, and a plot, and characters, and motives, and all the components of a proper story so that they can pull all the information they have together into a cohesive narrative that defines reality for themselves and others (Boje, 2001).

This unfolding, emerging narrative process happens with all of us when we experience something out of the ordinary that creates dissonance in our perceptions of how the world is supposed to work. We move on but the dissonance remains, lingering somewhere in our psyche. We think about the incident and try to make sense of it. Sometimes we are able to rapidly make sense thereby resolving the dissonance, but it may take an extended period of time to resolve the dissonance and engaging in a narrative helps the sense making process. For example, something happens at work that we are trying to make sense of, we come home and say to our spouse or a friend,

"The craziest thing happened today at work…" and proceed to describe what happened, concluding with, "What do you think?" The listener might then say, "Well, perhaps so and so, said thus and such, because…" and picks up the sense making where we left off. Through this process of give and take, sense is made, sense is given and a complete story is formed with a conclusion, at least for that iteration of the sense making process. The process continues as new information is received to either reinforce the narrative as it is, or modify it, perhaps creating a whole new story.

Some anecdotes, parts of stories, and parts of the dialogue surrounding the stories, are used in the following chapters to describe the leadership narratives found in the stories of Navy Chiefs and interpret their meaning. (Table 3 illustrates the frequency of the leadership narratives and their prescribed practices at the Senior Enlisted Academy and on the Ship.)

Chapter 7

YOU ARE PART OF SOMETHING BIGGER THAN YOURSELF; THEREFORE TAKE CARE OF YOUR PEOPLE, SUCK IT UP AND TRUST

You gotta trust that the guys on duty are doing their job, and the guys not on duty can sleep soundly knowing that you are doing your job.

Senior Chief Jerry

The leadership narrative *You are part of something bigger than yourself* was reflected in many of the stories collected; it was expressed through phrases like: "You're part of a higher calling," "We're all in this together," and "You are part of something bigger." Many of the stories that reflected this leadership narrative emphasized the importance of commitment to something beyond the individual, such as a greater good, or a purpose that transcended the individual and the immediate environment. They emphasized a oneness with the organization or the team. They emphasized supporting the team and recognition of team accomplishments over individual accomplishments. They emphasized sharing in the glory of a job well done. They also implied, or overtly stated, that individuals are expected to subjugate their personal needs to the needs of the organization, writ large or small; therefore members of the organization need to trust the process, trust others, or trust the system.

The leadership narrative *You are part of something bigger than yourself* resulted in three practices that repeated in the stories: 1) *take care of your people*, 2) *suck it up,* and 3) *trust.* These leadership practices at times worked together, but at other times presented a duality that the Chiefs wrestled with through their stories and in the discussion surrounding their stories.

TAKE CARE OF YOUR PEOPLE

The first practice that resulted from the leadership narrative *You are part of something bigger than yourself* was *take care of your people.* The practice *take care of your people* was expressed through phrases like "Take care of your people," "Take care of your guys," "Take care of your shipmates," and "Take care of your brothers and sisters in the Mess." It emphasized acting in ways that nourished the whole of the group or the organization. *Take care of your people* implied a mutuality among, and between, individuals and the group, or the team, or the organization. It also implied that the organized whole is more than the sum of its parts.

The leadership practice *take care of your people* has become a trite platitude in recent years. Trite because leaders and managers often say that taking care of people is a top priority while their actions tell a different story. The stories that travel through organizations have an exceptionally facile way of exposing trite platitudes and organizational falsehoods. Similarly, stories by their very nature, support the authenticity and integrity of organizational ideals. The Navy Chiefs gave numerous concrete examples of how they, or others, in the organization "take care" of people. Senior Chief Tim described risking his life for a shipmate in *USS Kitty Hawk Mishap*. In *Page 10's* Master Chief Tom described how one of his mentors took him under his wing. In *Lorelei Goes Home* Chief Sara showed sensitivity and respect through her care of a dead young woman. Senior Chief Evers described his angst when he was not able to save "his guys" in *Jet Blast Deflectors Malfunction*.

In the following romantic epic, Senior Chief Dee, a logistics specialist in her 40s, described the intervention of a hero, a mentor whom she admired, who helped her, and her shipmates, prevail and become successful. The mentor created a collaborative environment and a commitment to something "bigger than yourself."

NEW CHIEF BRINGS HOPE

I was basically stuck as a 2nd class petty officer. I was really frustrated. I couldn't seem to advance and I thought I was going to get kicked out of the Navy. But then in walks our new Chief. She had made Chief in nine years and I was like "Wow, super woman." And she was like a real shy kind of person. In her shy way she got to know everyone. Morale was bad in the division when she came. There were little cliques and everybody was always complaining about one thing or another. And she came in and got us studying so that we would make rate and started making studying fun. And she got us involved in community service. We did Meals on Wheels and stuff like that. And it was just great. And all of sudden we were like a team and there was camaraderie. Thanks to her, we all made rank, every single one of us. I thought I was going to retire as a 2nd class and here I am today a Senior Chief. She had an impact on everyone who was there. She brought out the best in us. She nominated me for sailor of year and I thought, "Wow, I never dreamed that I could be sailor of the year" and I actually got it. And she got an award when it was time for her to transfer and she read it to the division, she said, "this is not my award, this is your award, this is what you [the division] did." We still keep in touch and she is doing really well—she went LDO [Limited Duty Officer]—she's a lieutenant now. I would have gotten out of the Navy if it wasn't for her.

This is not an extraordinary story, many of us can relate to it. We have worked in organizations that were energized by the introduction of a new supervisor. In this story, Senior Chief Dee described a leader who created an atmosphere of trust that invigorated the workplace, fostering collaboration and enhancing energy flow within the division. Senior Chief Dee said "We all made rank, every single one of us," evidence that the entire "team" benefited from the actions of the new supervisor. Senior Chief Dee was using the story of her mentor to make sense of what happened to her, and she was using the leadership narrative *You are part of something bigger than yourself* to inform a practice—*take care of your people.* Senior Chief Dee implied that the "care giving" in the group was mutual and that by taking care of others one was taking care of oneself. After telling this story Senior Chief Dee described other incidents in which she shared in the glory of a job well done. The story provides a behavioral map for how to *take care of your people.* There is every reason to believe that Senior Chief Dee will use the leadership practice *take care of your people* with the people she will supervise in the future, and they in turn will model that behavior for others.

SUCK IT UP

The second practice resulting from the leadership narrative *You are part of something bigger than yourself* was *suck it up.* It was expressed through phrases like "suck it up," which often entailed "not whining" or "not being a cry baby" and subjugating individual desires, and ego, ostensibly for the greater good of the team. The leadership practice of *suck it up* reflected the need to subjugate personal needs to the needs of the organization. At times the practice *suck it up* worked with *take care of your people* and *trust,* creating an environment in which the needs of the individual are secondary to the needs of the team or the larger organization, but there was *trust* in the mutuality of that arrangement. The leadership narrative *You are part of something bigger than yourself* combined with its prescribed practices of *take care of your people, suck it up* and *trust,* implied that if you are one with the organization, or the group, it will be one with you and that relationship will be mutually beneficial.

Life and work onboard a warship is physically, mentally and emotionally demanding. Sailors live and work in cramped, industrial like, quarters apart from loved ones, for weeks, or months at a time. The workday is long. Personal freedoms are severely restricted and leisure outlets are limited while at sea.

A common response to complaints about habitability or living conditions at sea is, "this is a warship not a cruise ship," implying *suck it up*. While Navy pay and benefits are not bad, they are not great, given the personal sacrifice expected and required. Yet, despite the hardships there was the sense that job satisfaction somehow makes the personal sacrifices worthwhile.

Most Navy Chiefs as individuals have decided to make the Navy a career so it makes sense that they would feel that the advantages outweighed the disadvantages of their chosen careers. Individual members of the Chiefs' Mess often used the phrase "Every day at sea is a good day."

This phrase was used to communicate different meaning, in different places at different times. The Chiefs used "Every day at sea is a good day" in a sarcastic manner to chide and motivate subordinates who were engaged in work that "sucks." They used the mantra "Every day at sea is a good day," to reinforce their bonds as disciplined professionals who only truly engage in their craft when they are at sea. And they used it to communicate that it is easier to keep the young sailors who work for them out of trouble and focused on their work when they are at sea. In general the Navy Chiefs were more outwardly positive than negative but in their most candid moments there were those who admitted that at times life at sea and the work "sucks." But there was the shared perception that while Navy life could be demanding it also could be intensely satisfying.

The following epic-comedy illustrates the leadership narrative *You are part of something bigger than yourself* and all three of its prescribed practices; *take care of your people, suck it up* and *trust*. Senior Chief Sam, the narrator and protagonist in this story, is a large, soft-spoken Aviation Boatswain Mate with a shy smile who still looks like the football linebacker he was in high school. Senior Chief Sam was in charge of the Arresting Gear Division on the Ship— the division that maintains the thick metal cables and machinery that stops, or "arrests," the aircraft when they land on the flight deck. He told the following story while supervising several sailors doing maintenance on the equipment during a pause in flight deck operations. The Arresting Gear Machine Compartment, one of several low-ceilinged rooms located directly below the flight deck, smelled like hydraulic fluid on hot metal and the entire room shuddered when jets slammed into the deck and hooked the cables. It was hot and it was impossible to carry on a conversation during flight operations due to the loud whirring of machines, cables and jet engines. The work is physically demanding. In this story Senior Chief Sam described how he came to join the Navy and how he became an Aviation Boatswain Mate—an "AB".

AB OR WANNA BE

The recruiter said "So when can you be ready to go?" And I said "I'm not doing anything else right now, so I guess I could be ready to go tomorrow." Well the recruiter's eyes lit up when I said that and he said "Could you excuse me a minute while I make a phone call?" And he goes off and makes a phone call and comes back and says, "You're in luck. We just happen to have an opening and you can leave tomorrow." I was a little nervous then and I said, "Well, what will I be doing?" And the recruiter said "I suggest you go 'undesignated' [no extra schooling for a technical specialty] then when you get out to the fleet you can shop around, try out different jobs, and see which one you like best." I thought that sounded good, so I signed on and shipped out for Boot Camp [basic training] the next day. I made it through Boot Camp and was sent to a carrier [aircraft carrier] and when I arrived at the carrier, there were about eight of us that checked in that day, four of the guys were sent off to various departments because they were "designated" [they had already been trained for technical specialties] and the four of us who were "undesignated" were sent to deck [Deck Department]—from there I was sent to work with the arresting gear. It was hot. It was hard. It sucked! So the next day I thought, "I'm not goin' back there—I'm gonna 'shop around' just like my recruiter said— I'll go work in the ship's Post Office instead." So I went down to the ship's Post Office and the chief there says, "What are you doing here?" And I said, "Well, I was working with the arresting gear yesterday and I know I don't want to do that, so I thought I would come down and try working here." And the Chief gives me this [disgusted] look and says, "Now look son, that's not how it works. You get your ass back up to Arresting Gear. I'll let the Master Chief know you are comin.'" When I got back up to Arresting [the Arresting Gear Machine Compartment] the Master Chief was waiting for me and he says, "Why didn't you report for work today?" And I said "With all due respect Master Chief, I did not like the work much yesterday and my recruiter told me that if I was "undesignated" I could shop around and try different jobs until I found one I liked. That old Master Chief looked like he was about to explode, and he says, "Son, there are only two ratings in the Navy, AB [Aviation Boatswain Mate] and Wanna Be, now which are you gonna be an AB or a Wanna Be?" And I said "I guess I wanna be an AB, Master Chief" and he said, "Right answer! Now get to work!" And here I am today.

This story, like *French Fries,* had a gallows humor quality. Outsiders might see this story as a "tragedy" because it appears that Senior Chief Sam was the pawn of a trickster—a recruiter who was out to fill a quota. But this story is humorous to insiders because almost everyone in the military has had a trickster experience with a recruiter or detailer (the human resources specialists who put service members into jobs), or they know someone who has had such an experience. Since the days when sailors were regularly pressed into service in seaport taverns, the Navy has needed undesignated sailors to work the deck of warships at sea—work that was learned from the stories and examples of the "old salts," not textbooks.

The day that Senior Chief Sam walked into the recruiting office, the recruiter saw an opportunity to make a quick sale and fill a quota. He took advantage of Senior Chief Sam's ability to leave for Boot Camp the next day. The recruiter knew that it was unlikely that Senior Chief Sam would be able to "shop around" for a job once he got out to the fleet, he knew that the work of an undesignated sailor is not glamorous. The recruiter did not have Senior Chief Sam's future job satisfaction foremost in mind the day he sent him off to Boot Camp, but he also was not being totally disingenuous. The recruiter was a career Navy man himself so he believed in the organization. The dominant narrative of *You are part of something bigger than yourself* was probably part of the recruiter's make up or he would not have chosen a Navy career and would not be a recruiter. In his way he was *trusting* that Senior Chief Sam would learn to *suck it up,* adapt and be successful.

Many who went before Senior Chief Sam adapted to the organization and many who came after him would as well. The ones who failed to *suck it up* and *trust* the system would either get kicked out, or would choose to leave the Navy after their first tour of duty. Senior Chief Sam resisted when he first arrived at the ship but then he decided to *suck it up* and he was *taken care of* by the gruff, but attentive, Chiefs who taught him his trade and provided him with a model for how to lead others. So although Senior Chief Sam was "shanghaied", the story had a happy ending—he has had a successful career and is obviously proud of his technical specialty today. He used this story to make sense and he also used the story to describe some of the qualities needed to thrive in the organization, such as a sense of humor and a willingness to conform.

There was a young sailor doing maintenance on a cable nearby while Senior Chief Sam was telling this story; when asked if he had ever heard the story before he laughed and replied, "Oh yeah!" Senior Chief Sam's subordinates had heard this story before, it no doubt helped them make sense and helped them feel proud of what they do as well. Senior Chief Sam expects the people who work for him to *suck it up* on days when the planes are flying around the clock, yet he also expressed his desire to *take care of his people* even while they were *sucking it up*. In the following excerpt from the discussion that followed the telling of *AB or Wanna Be*, Senior Chief Sam was negotiating common ground between competing leadership narratives and practices.

SENIOR CHIEF SAM CONTINUED...

One of my biggest challenges is making sure the guys are getting what they need to do the job. During flight ops [operations], depending on conditions, we might not be able to stop for a hot meal [during a 12 hour shift]. This is hard work and lots can go wrong. If we can't break the guys loose to go down to chow I make sure that we at least get something brought up here from the galley. It might not be a hot meal but I make sure they get something and we make sure that they get a break if they need one. You need to be able to read people in this job. You need to know your guys. You need to be able to see it when someone needs a break or someone is going to get hurt.

At times open contradictions and dualities between leadership narratives and practices were apparent. Like parents who lament the insolance of their children, some of the Chiefs expressed frustration that they sometimes feel forced to *take care of people* when the people being taken care of should be able to *suck it up*. Master Chief Jeff commented, "We are expected to coddle these kids too much nowadays." Senior Chief Tray alluded to the duality between allowing people to make mistakes and taking care of them when he commented,

A lot of Chiefs resent the baby-sitting they have to do today. This 'zero mistake mentality' is ridiculous. Eighteen to 20 year olds are going to make mistakes and there is just not much you can do about it. At least when we are at sea they are contained but when we are in port you cannot control everything.

Through the interplay of leadership narratives and practices, Senior Chief Sam described how he was coping with the dualities and dissonance in his environment, and discovering actions that he and the group must take to prosper. He was communicating that there is a balance between *taking care of your people* and expecting subordinates to *suck it up*. He was saying that he did not want "his people" to make mistakes, because in his line of work mistakes can be fatal. He was saying that he could expect them to only *suck it up* so much before they break down.

TRUST

The need for *trust* emerged from many of the Chiefs' stories and narratives. It was expressed in phrases like "Sometimes you got to trust that the guys making the decisions know what they are doing," or "Sometimes you don't have time to ask questions or get into explanations," or "Sometimes you need to trust your gut," and "You've got to trust that the people on duty are doing their job." Other phrases and key words that reflected this leadership practice were: "trust your gut," "trust your shipmates," "trust the system," and "rules, or SOPs (Standard Operating Procedures) are written in blood."

In the following story, Senior Chief Jerry, the Chief in charge of the Electronics Warfare Division, referred to the "Stark incident"—an incident that took place during the Iran-Iraq war in 1987. In 1987, an Iraqi fighter jet fired two missiles at the USS Stark, a Navy warship. The missiles hit the ship. Thirty-seven sailors were killed and 21 were injured in the incident.

Senior Chief Jerry was the Chief in charge of the division that monitored the radar equipment on the Ship that was designed to detect enemy ships, enemy aircraft, and enemy weapons that might be a threat to the ship. Therefore, the Stark incident had special relevance for the sailors who worked in Electronics Warfare. Senior Chief Jerry, walked fast, talked fast, and had a ready wit. The following discussion took place in the compartment on the ship where the air traffic controllers and other radar specialists work. The compartment was rather dark—illuminated mostly by fluorescent screens of the electronic scopes that lined the room. Approximately eight sailors sat at various stations around the room monitoring the flashing blips on the scopes as Senior Chief Jerry admonished them to "Remember the Stark".

REMEMBER THE STARK

I am the old Yoda here. Most of these kids are too young to remember incidents like the Stark—they were babies in 1987—ha—some of them weren't even born yet. So I tell them these stories over and over again. I say remember what happened on the Stark. I was not on the Stark, but I will always remember what happened and I want these guys to remember so they don't make the same mistakes. The Stark didn't get off a single shot in its defense. They did not pick up those missiles on radar—that should not have happened. The guys who were killed were the ones who were asleep in their racks in berthing, not the guys on duty. So I tell these guys—your shipmates are depending upon you to do your job. If you have any questions about what you see on the screen, look it up, ask someone else to take a look, trust your gut, if something doesn't look right check it out.

Although *Remember the Stark* was not a proper story by itself it contained a reference to a proper story—the Stark Incident—a story that was a tragedy. The sailors who listened to Senior Chief Jerry's admonition to *Remember the Stark* said that they were familiar with the Stark incident and other tragic incidents they studied, even though "they were babies, or not even born" when many of the incidents occurred.

The observation and dialogue surrounding *Remember the Stark* reflected dualities between leadership narratives and practices as Senior Chief Jerry negotiated behavioral maps he—and the people in the storied space of the Electronics Warfare Division—must follow in order to survive. He was saying "trust your gut" but check your perceptions with someone else. He was telling the sailors who worked for him that their shipmates *trust* that they are doing their job. He was telling the sailors who work for him that one small blip on a screen that they miss, or that they do not check out, can result in disaster and loss of lives.

Another reoccurring phrase that reflected the leadership practice *trust* was "SOP's (Standard Operating Procedures) are written in blood." Many of the Chiefs used this phrase. Senior Chief Rolf, a Boatswain Mate in the Deck Department on the Ship explained the need to follow Standard Operating Procedures and rules this way,

Every one of the SOP's [Standard Operating Procedures] we have has a story behind it—an incident that happened. Somebody lost an arm, or a leg, or was killed. That is why the SOP was written. I try to explain this to these young sailors when I have time. I try to tell them about an incident I know about, or a personal experience I had, or a reason, but sometimes I am not there, or there is not time and they just need to follow the SOP.

The implication was that Standard Operating Procedures and rules help individuals and organizations survive in a sometimes chaotic, dangerous environment. But even though "SOP's are written in blood" they apparently are not etched in stone. Senior Chief Rolf explained that Standard Operating Procedures are often changed or modified in response to new information, he said, "We are always getting new SOP's and some change. Something happens. Someone gets hurt and we get a new SOP." While emerging incidents result in new standard operating procedures or changes to old standard operating procedures, the leadership practice *trust* in this case "trust the system" has prevailed.

The leadership practice of *trust* often occurred in tandem with the leadership practice *suck it up*. Master Chief Cal, the Master Chief in charge of the Air Department, expressed the need to *trust* as follows,

> *There are times when you just need to do what you are told—there just isn't time to ask questions—you need to do what you are told and do it now. You need to trust that the people you are working for know something you don't, or have information you don't have—there's just no time to ask questions.*

In an interesting twist on this leadership practice, Master Chief Hassan, one of the instructors at the Senior Enlisted Academy, cautioned, "When something doesn't feel right you need to trust your gut. You need to have the courage to ask the hard questions." In the following cautionary tale Master Chief Hassan described a disciplinary system breakdown that threatened to erode the trust of the crew. Master Chief Hassan, who also told *French Fries*, has a low raspy voice as well as the timing and delivery of a professional comedian—everyone paid attention when he spoke.

DUDE, WHAT HAPPENED?

I was on a Spruance Class Destroyer [a medium sized warship] and I was walkin' by the Master at Arms office [on board security force] one day and I see this kid with his arm wrapped in gauze and I says "Dude, what happened to you?" and he says "Do you really want to know what happened Chief? Because no one else seems to want to know what really happened." And I says, "Sure, tell me what happened." And he explains that he was in the chow line and he was grab assin' with one of his buddies and in the process he bumped into the tray of the female sailor in front of him and her orange juice spilled, and she came unglued and turned around and threw a cup of hot coffee on him, burning his arm. And now he was on report [facing disciplinary action] for starting a fight. And I am thinkin' to myself there are two sides to every story and I am sure I will hear the other side of this story at a Disciplinary Review Board [a disciplinary review at which the Chiefs decide if a minor incident should go to Captain's Mast] and it will get sorted out there. So I go about my business and the next day I find out this incident isn't going to Disciplinary Review Board—this kid went straight to Captain's Mast and the Captain threw the book at him. So I go to the Command Master Chief and I say, "What's up with this? Why didn't this incident even go to Disciplinary Review Board?" And the Master Chief says, "Your young friend was messin' with the wrong girl." Well, I come to find out that the girl who threw the cup of coffee was the Executive Officer's yeoman and a shoo-in for Sailor of the Year. So I tell the Command Master Chief "This just ain't right. This shouldn't have gone to mast. Nobody even listened to Seaman Roger's side of the story. There were other people who saw this. They're not stupid. What kind of a message is this going to send to the crew—that if you're the XO's yeoman you can get away with whatever shit you want?" I was a brand new Chief at the time and I was just disgusted with the whole Mess [Chiefs' Mess]. I thought I don't know if I even want to be a part of this group of pussies. I got up at the next Chiefs' meeting and I told them it was a travesty. Sometimes somebody needs to stand up and ask the hard questions. We blew our trust with the crew on that one and it would be hard to get it back.

In *Dude What Happened?* Master Chief Hassan described a system breakdown. He described losing the *trust* of the crew due to a perceived injustice. Master Chief Hassan used this story to illustrate a leadership practice; he suggested that sometimes you need to challenge the members of your "storied space" to do what is right—even when it means going against the grain. Through this story, Master Chief Hassan implied that *trust* is reciprocal. He explained that sailors need to feel that they can *trust* their supervisors and vice versa for the system to work. In this story Master Chief Hassan was making sense and giving sense. In the dialogue surrounding this story he was negotiating a delicate balance between leadership narratives and practices. He did not say what it took to win back the trust of the crew—perhaps in this case the leadership never did.

UNITY, INTERCONNECTEDNESS AND
THE ABILITY TO SELF ORGANIZE

Many of the Chiefs told stories that reflected tremendous pride, patriotism and commitment to the organization's mission and the Navy's mission. The missions were perceived as noble. And this commitment and *trust* in *something bigger than yourself* enabled the protagonists in the stories to cope when they needed to *suck it up*. Indeed, the leadership narrative of *You are part of something bigger than yourself* seemed to be a major contributor to job satisfaction despite the fact that it entailed personal sacrifice without a great deal of material gain.

A thread that runs through the writing about Complexity Science—the theories that inform much of the interpretation in this book—is that there is a unity and interconnectedness between all living things and beyond to an interconnectedness with the universe. This is not a particularly revelatory concept, it is present in many Eastern and Western religions that comprise the cultural foundations of organizations around the world. But a Complexity Science view of how living systems function explains that unity and interconnectedness prompt entities to self organize at the lowest possible levels and expand outward in repeating patterns that may not be exactly the same but are very similar. The notion of unity and interconnectedness challenges a top down view of how the universe works, suggesting that the seeds of emergence and creation are within living systems and suggests that the unifying force within the universe may just as likely be bottom up as top down.

I do not intend to get into the theological underpinnings of organizational cultures—that is another book and another body of research—but I can say that there was evidence of unity and interconnectedness in the stories and narratives of Navy Chiefs. There was evidence in the stories of a willingness to *suck it up, take care of your people* and *trust* even when doing so was not in the self-interest of the individual. There was evidence of unity and interconnectedness in the altruism described in the narratives, even when the people being taken care of were strangers, and models from natural systems propose at least one explanation for altruism.

The narrative *You are part of something bigger than yourself* underscored the principle of patriotism sometimes referred to as political unity (Gell-Mann,

1994). Gell-Mann (1994) explained that while protecting others, such as family members and even extended family members, makes evolutionary biological sense, protecting others beyond family may also make sense for humans. He said,

Sociobiologists now agree that patterns of altruistic behavior in humans are greatly affected by culture. A certain willingness to risk one's life for another human being can easily extend to all members of one's tribe ... on the scale of a nation state, it is known as patriotism. As people have aggregated into larger and larger societies the concept of 'us' has tended to grow in scope… community is essential to human activity, but only communities motivated to work together are likely to be adaptive in the world of the future (Gell-Mann, 1994: 360-361).

The storied spaces of the Ship, the Chiefs' Mess and the Navy displayed evidence of community in the leadership narrative *You are part of something bigger than yourself* and its attendant practices *take care of your people, suck it up* and *trust*. People in the storied spaces studied were conditioned to take care of one another. Some stories, like *USS Kitty Hawk Mishap* described altruistic behavior. Other stories described a willingness to subjugate personal needs to the needs of the group.

CHAPTER SUMMARY

The leadership narrative *You are part of something bigger than yourself* prescribed at least three leadership practices (*take care of your people, suck it up* and *trust*) that worked alone, or together, to create behavioral maps for the Navy Chiefs described here. Through the discourse surrounding the stories the Chiefs tested the flexibility of the narrative *You are part of something bigger than yourself* and its attendant practices, negotiating ways to apply them in different situations.

The group's use of this leadership narrative and its practices enabled the group to function like a Complex Adaptive Entity—a self organizing system that sustains itself. The leadership narrative *You are part of something bigger than yourself* implied that indeed the whole is greater than the sum of its parts. It implied that individual ego should be subjugated to the greater good. But it also implied a mutuality, a complementary system in which individuals act in ways that support the greater good and in turn they are taken care of. The stories and the discourse surrounding them indicated that at times leadership practices contradict each other, but when practices are contradictory balance can be negotiated until the contradictions are resolved.

Sense making took place as the Navy Chiefs told their stories—narratives that prescribed ways of responding to the challenges they face each day. Through the dynamic process of crafting narratives that define the way they view the world the Chiefs were able to negotiate a balance between leadership practices, at times expecting their people to *suck it up* and at other times *taking care of their people*. Sometimes the Chiefs themselves needed to *suck it up* and they were able to cope with their own personal sacrifices because they trusted the system, trusted each other and trusted the people they worked for. The narrative *You are part of something bigger than yourself* helped define their view of reality. It was a commonly accepted narrative that helped to unite the organization.

LEADERSHIP REFLECTION

New views of evolutionary biology suggest that collaboration has been just as important—perhaps even more important—than competition in the explosive growth of the human species. When people feel they are part of a higher calling, "something bigger than themselves," they are willing to sacrifice to support their group, or their organization.

1. *Is there the feeling in your organization that there is a commitment to something beyond, individual self-interest? Do you as a leader feel committed to something beyond your own self-interest? What stories in the organization support or challenge that feeling?*

2. *When there is trust in an organization, there is a greater willingness to collaborate, work as a team and creatively share ideas—resulting in positive energy and synergy. Are there stories in your organization that reflect a sense of trust? Are there stories about collaboration and teamwork, or do stories predominate about personal gain?*

3. *When people feel a sense of satisfaction in their work, and the organization they are a part of, they are willing to make personal sacrifices for the greater good. Are there stories in your organization about people who were willing to make personal sacrifices to support the team? What message do these stories communicate to the group? How can you use your stories and the stories of others in the organization to communicate the values you aspire to?*

4. *There is a delicate balance in all organizations between taking care of people, expecting them to sacrifice and expecting them to trust—how do you as a leader manage that balance? What sacrifice and rewards stories predominate in your organization and what values do they project? What are the stories in your organization about motivation?*

5. *Biological systems provide models of self-organization from the bottom up. What stories circulate in your organization about initiative, trust and risk taking? What attitudes do those stories reflect? What behavioral maps do those stories prescribe?*

6. *What themes connect the stories you identified above and what are the dominant narratives? Dominant narratives that may be at play: "You are part of something bigger than yourself," "The ends justify the means," "Every man for himself."*

(See Appendix D (Storytelling guidelines) to hone storytelling skills)

Chapter 8

SH*T HAPPENS: THEREFORE, LIVE AND LEARN, KEEP YOUR HEAD ON A SWIVEL AND "SEMPER GUMBY" (STAY FLEXIBLE)

Ass chewings are free—I call it growth through uncomfortability

Master Chief Cal

The leadership narrative *Sh*t happens,* and its attendant leadership practices *live and learn, keep your head on a swivel* and *"Semper Gumby"(stay flexible),* reflect the theme that despite the best laid plans the unexpected will happen and mistakes will be made, therefore you need to learn from your mistakes, always have your wits about you and remain flexible so that you can adjust to change that is seen as inevitable. This leadership narrative and its attendant practices will be examined one at a time.

LIVE AND LEARN

Live and learn was the leadership practice most often repeated by Navy Chiefs. This leadership practice prescribed a way of responding to information, the environment, subordinates and each other in the storied spaces of the Navy, the Chiefs' Mess, the Ship, and the Senior Enlisted Academy.

The following personal epic is an example of a story that reflected the leadership practice *live and learn.* In this story Master Chief Cal described how he came to be where he is today. Master Chief Cal was the Air Department Master Chief. The Air Department is the largest department on the Ship with approximately 700 sailors.

MASTER CHIEF CAL'S STORY

There is no good reason why I should be where I am today. Most of the guys from my neighborhood are dead or in prison. I figured the Navy was my way out. I got into a lot of trouble early on and practically got kicked out a couple of times. I went to captain's mast [non-judicial disciplinary proceedings] for fighting, alcohol, drugs, you name it. The captain would throw the book at me—I was restricted and busted [reduced in rank]. I got my ass chewed plenty. But when I was at work I was okay, I worked hard and did my job—that is what saved me. My problem was that I got into trouble when I was on the beach [off the ship]. I was a hard head. It was painful but eventually I learned. So I tell these kids today "Ass chewins' are free"—it's a free lesson right there and you don't pay nothin' for it—learn from it. But learn from it before I did, because the way I learned is just too damn painful! Even hard heads like me can learn, I call it "Growth through uncomfortability."

In this story Master Chief Cal explained that it is acceptable to make mistakes because they are an integral component of growth. He was reflecting his belief that life and learning are often a struggle and that the struggle needs to be embraced rather than avoided if growth is to take place. Master Chief Cal used this story to make sense of how a young man who grew up in a rough neighborhood, and made many mistakes as a young sailor, could succeed in the Navy. He used this story to make sense of how he survived while others did not. The leadership practice *live and learn* combined with other leadership practices such as *suck it up*, and *take care of your people* provide Master Chief Cal with a way to respond when he receives information about others who have made mistakes, or when he makes a mistake himself.

All of the stories Master Chief Cal told contained this leadership practice. As the Master Chief responsible for the largest department on the Ship, Master Chief Cal was constantly dealing with mistakes that people within the department made. It was clear that Master Chief Cal was regularly balancing the practices of *live and learn, take care of your people and suck it up* as he dialogued with others on the Ship. His testing of these practices reflected the daily struggle of life and work in his environment.

Some Chiefs, such as Master Chief Cal, emphasized *live and learn* more than others. The Chiefs on the Ship repeated this leadership practice more frequently than the Chiefs at the Senior Enlisted Academy. Some of the other phrases used to express this narrative were: "People make mistakes so smack yourself and get back to work," "You can grow out of being a screw up," and "18-20 year olds are going to make mistakes and there's just not much you can do about it."

In the following story, and narrative surrounding it, Master Chief Cal discussed how the Navy has changed over the years. In this story and emerging narrative he was testing the leadership practice *live and learn*. He suggested that sometimes people in the Navy do not have the luxury of learning from their mistakes. He suggested that while it is acceptable to make mistakes, it is not acceptable to repeat the same mistakes. He also acknowledged that some mistakes are unforgivable and he described feeling somewhat conflicted about that contradiction. In this emerging narrative Master Chief Cal attempted to resolve the dissonance he perceived between what the Navy was like when he was a young sailor and what it is like today.

MASTER CHIEF CAL'S STORY CONTINUED...

I don't think I would make it in the Navy today. Two alcohol incidents and you get kicked out. One drug offense and you get kicked out. Sex on the ship is a court martial offense. One domestic incident and you are in trouble. The climate is tougher today. And it is hard to get through to these kids. I love it when we are at sea—at least you have a fighting chance of controlling their behavior—but it is almost impossible when they [the sailors] are on the beach [off the ship]. We [the Chief Petty Officers] try to get through to them before they get into trouble we cannot get them out of, but sometimes it seems like we are fighting a losing battle.

Master Chief Cal was wrestling with his own dissonance in the previous emerging narrative—the dissonance between *taking care of your people* and the belief that some mistakes are unacceptable. He was expressing the struggle he faces daily as a leader, mentor and shipmate. In the following story Master Chief Cal gave an example of how the environment has changed since he was a young sailor.

MJ IN THE SALSA

We had this kid at mast the other day who popped positive [random urinanalysis test for drug use] for marijuana. It was his second positive for marijuana. The first time he got off [was acquitted] because he had his wife and her mother come in and tell the Captain that they put marijuana in the salsa. The Captain bought that explanation and let him off. But the "salsa defense" only works once. He popped positive again and he was history—no questions asked.

Master Chief Cal was not defending the young sailor who "popped positive for marijuana." He was affirming his support for the Navy's "zero tolerance" policy for illegal drug use, while he wondered if he could even "make it in the Navy today" given such policies. He expressed *trust* in the Navy's random urinalysis testing program, but he also suggested that there are exceptions to the "zero tolerance" rule. Through his stories and the narratives surrounding them, Master Chief Cal was negotiating ways of adapting to an environment that has changed since he was a young sailor, an environment that is constantly changing.

Master Chief Cal also referred to "controlling behavior" and how it is easier to "control sailors' behavior" when the Ship is at sea. He expressed frustration that the behavior of the young sailors is beyond his control when the Ship was not at sea. It is important to remember that unlike most other organizations the Navy holds its members accountable for their activities twenty four hours a day, seven days a week, when they are at work and when they are off duty on their own time. This lack of control troubled Master Chief Cal as a supervisor. There is a tendency—perhaps especially in command and control type organizations—to want to try to exert control as the organization experiences imbalance and dissonance. But as Master Chief Cal recognized, growth often comes from struggle—what he calls "uncomfortability." The interplay between stories and emerging narratives as people and organizations attempt to cope with dissonance and adapt to ever changing environments can enhance individual and organizational creativity and uncover new and more effective paths to adaptability.

The lesson from natural systems is that phases of dissonance, struggle and even chaos are normal and indeed a necessary component of growth. Story, and the narrative discourse that accompanies it, sometimes reflects dissonance and the struggle to learn that is integral to the adaptability of an entity. Stories and narratives can also help individuals and organizations identify ways of behaving—practices—that are maladaptive or in need of modification. While this may seem like common sense and many organizations claim that they allow individuals to learn from their mistakes, in practice most organizations punish individuals for making mistakes and discourage risk taking that might result in growth.

The following story illustrated another aspect of the leadership practice *live and learn*. Just as individuals and organizations can learn through the tragic stories of others, they can learn through the mistakes of others. In the following story Chief Dante, an engineering specialist, who was a student at the Senior Enlisted Academy, described an uncomfortable incident that took place on a deployed submarine. Chief Dante survived this incident, which made this account a personal epic.

BOW PLANES INCIDENT

*I was on a Sea Wolf [a type of submarine] and even though the class [this type of submarine] has been out for 12 or 13 years, we were still working the bugs out. And we were having problems with the bow planes [large retractable stabilizers that are used to help maneuver the submarine]. The automatic system just wasn't workin'. So the Captain and XO [Executive Officer], the engineer and myself gets together and we decided that we would work the bow planes manually until we work the bugs out and since I had the "A" gang [sailors who work in auxiliary engineering], and I was the one dealing with all the hydraulics, and I knew the electrical system pretty good, I would be the one to manually operate the bow planes with a couple of sailors until we worked the bugs out. But then one day the Captain comes into the control room and he wants to retract the bow planes, so he says to the OOD [Officer of the Deck], "We gonna retract the bow planes" so the OOD tells the messenger to go get me. It's a submarine so he needs to find me, even though submarines are not big he doesn't know where I am, so it's gonna take a couple minutes. So two minutes later the Captain comes back into the control room and he is getting angry and he says, "I said retract the bow planes!" And the OOD says "Sir, they're looking for Dante now, he should be here in a minute" and the Captain is like "This is my fucking ship, I said retract the bow planes, DO IT NOW!" And so the OOD this JG [lieutenant junior grade] says "Yes sir!" and he hits the button on the panel to retract the bow planes and at that moment I walk in and I see what's goin' on and I says "What? You guys retracted the bow planes?" and the OD looks at me, and the Captain looks at me, and I say "Are you crazy?" And everybody looks at me, expecting the Captain to chew me out because I jus' said "Are you crazy?" so I went over to the panel to where the controls for all the hydraulics are to try to stop it and as I am walkin' over to the control panel I hear this big BOOM and the whole submarine shudders and what happens is that the bow planes went part way and got stuck. And so now I can jus' picture what has happened—this big piece of equipment jus' got tore up—there are gonna be big damages. And then the Captain is all desperate because he knows his ass is on the line and he says "Dante, what can we do?" And I says "Sir, the only thing we can do is try to unjam it manually." So the Captain says "Well, jus' do what you need to". So I try to unjam it but it is jus' not working. So I says "Sir, we can't do it, we are going to need to surface and we may need to pull into port." And he looks down and stamps his foot and he is like "F***! F***! F***! F***! F***!" and he went to his stateroom and he didn't* *come out until the next morning jus' before we were about to pull pier side. He didn't talk to no one. He was acting jus' like a little kid.*

This story resonated with the group and sparked a series of stories about leaders who were not willing to own up to their mistakes. Everyone in the group had a story about a leader who made life miserable for the people who worked for them and displayed behavior that the observer vowed never to repeat. Master Chief Jeff made the following observation in the discussion that followed the telling of *Bow Planes Incident*.

Ya know, it's interesting, we all have these stories about the assholes we have worked for, they seem to make an impression and if you survive you learn what you don't want to be like. But what about the guys who are just doin' their job, and gettin' the job done, day in and day out? You never hear much about them. What's up with that?

Indeed, when relating stories about leaders, the Chiefs tended to tell stories about those at opposite ends of the spectrum, those they admired and respected, or those they loathed. Interesting stories have a plot, that is to say a predicament that the protagonist must resolve. Therefore examples of poor leadership often contain the ingredients of memorable, compelling stories. It seems as humans we have a tendency to remember the people at opposite ends of the spectrum. The people we love or the people we hate are the ones we tell stories about. They are the epics, tragedies, comedies and romances of the work place. Stories like *Bow Planes Incident* are rich. They contain colorful characters, problems, conflicts, struggles, risks and danger. But they also contain a message—a moral—a prescription to not act this way. The Chiefs who told stories about struggling and surviving despite bad leadership— and the bad leaders were not always superior officers, they were sometimes fellow Chiefs—sometimes survived by biting their tongues and sucking it up. At other times they survived by sharing their struggle with others who were also suffering the abuse of the poor leader. Through sharing the stories about their suffering they created a bond with others and that bond helped them prevail. Like the crews in the classic stories *The Caine Mutiny* or *Mutiny on the Bounty* the Navy Chiefs coalesced around their views of what constituted bad leadership.

One of the female Master Chief Petty Officers on the Ship, Master Chief Quin, echoed the leadership practices of *suck it up, take care of your people,* and *live and learn*, in several of her stories. It was apparent in observations of her dealings with the sailors who worked for her that she expected them to conform, do their jobs and not complain, but she also implied that if they did so she would take care of them. Master Chief Quin, a tall woman with a commanding presence was in charge of the Human Resources Department on the Ship, told the following story,

RAISED BY WOLVES

I was part of the first wave of women to be sent to sea. When I was a young seaman, like all young seamen, I was sent to do a stint of "mess cranking" duty [the dirty work involved in food service and clean-up]. I had the good fortune of being sent to the Chiefs' Mess to do mess cranking. I don't think I would be where I am today if that hadn't happened. Those guys formed me. They taught me. They were tough but they were fair. You couldn't be thin skinned or wear your heart on your sleeve. But you also couldn't be around them and not pick up on what it meant to be a good leader. I learned enough about leadership in three months of mess cranking in the Chiefs' Mess to sustain me for an entire career. If you were willing to work hard and willing to learn, they took care of you.

Master Chief Quin echoed Master Chief Cal's comments when she said, "If you were willing to work hard and willing to learn they took care of you." This phrase, repeated by both Master Chief Cal and Master Chief Quin, described a proven practice that worked for them. It also reflected behavior they expected to see and would reward in their subordinates. This is not a revelatory or new leadership concept. But what happens when the members of an organization who have worked hard and expressed a desire to learn are "down-sized?" What happens when middle managers who have trained and mentored subordinates are told to "let them go?" *Trust* in the organization, and trust in the leaders in the organization, is violated when the leadership touts values that are not supported by the stories that circulate in the organization and are perceived as the truth. *Suck it up* ceases to make sense and members of the organization stop believing that *they are part of something bigger than themselves* when there are unreconciled differences between the organization's stated values and its practices. Fear and distrust rule as individuals look after their own interests and degradation of the entire organization follows when dissonance between values and principles persists. Calculated risk taking that can result in innovation and creativity is stifled as individuals strive to protect their own self interests and reevaluate their positions within the organization.

There were two other younger women—a Chief Petty Officer and a 2nd Class Petty Officer—present when Master Chief Quin told the following story. In *Single Parent on Sea Duty*, Master Chief Quin described how she had to *live and learn* while she had to *suck it up* to succeed in the Navy.

Single parent on sea duty

It hasn't been easy. Right after I made senior chief I was sent back to sea duty. I was divorced at the time and my son was young. I was stationed on the West Coast and my "ex" was on the East Coast. My son had been with me since he was born. He would spend some time in the summer with his father but most of the time he was with me. The Ship was working up for deployment. We were going to deploy in January. My son was going to stay with me until deployment and then go to the East Coast and stay with his dad while I was gone [six months]. It was September and already I was having trouble arranging for caregivers while we were out at sea for a week here and a week there, working up to deployment. My friends and neighbors were super supportive. They were always willing to take him in when I had to go to sea but I looked at the work up schedule and it was brutal—we were going to be gone a lot. It was breaking my heart to always have to leave my son with someone else while I was at sea. Finally I just had to face it, it was not a good situation for him and it was selfish of me. I wanted to keep him with me so much but I just couldn't do it. It broke my heart but I had to call up my ex and ask him if he could take him in September rather than January. It was so hard, but it was what I needed to do.

Tears welled up in Master Chief Quin's eyes as she told this story, and the young petty officer listening to the story became misty as well. Master Chief Quin was describing practices that she has used to adapt and thrive, and she was prescribing practices that the two younger women in the room would need to imitate if they wanted to succeed in the organization. When I asked the two younger women in the room about their plans for the future, one commented that she planned to get out of the Navy: "I want to have a family. My boyfriend is in the Navy and I just don't think I could do it [have family and stay in the Navy]. It would just be too hard." But the other young woman who had been listening to the story replied that she intended to stay on active duty and have a Navy career.

A few moments after Master Chief Quin told the two previous stories, a young, non-rated, male sailor, who was about to be discharged from the Navy and would soon leave the Ship, appeared at her door with his checkout sheet in hand for her signature. His work shirt was a bit wrinkled and his shoes were not shined. He had several small wooden pegs in his ears to keep his ear piercings from closing up (male sailors are not allowed to wear ear rings in uniform). He did not look at the Master Chief as he handed her his form to sign. She rose to her full five foot, ten inch height as she looked over the form and took a deep breath. Before signing the form and handing it back to him she said, "You understand that you've got 180 days to join the reserves after you get out and you will not lose your credit for time in service?" He mumbled that he understood. She then added in a stern voice, "Seaman Smith, for the next three days you are still in the Navy—look at me—while you are in uniform those pegs in your ears should not be visible." His face turned red, he mumbled something that could have been "Yes, ma'am," and quickly departed as soon as he had his signed form in hand. After he left Master Chief Quin let out a disgusted sigh and said,

There is a perfect example of a kid who is going nowhere, you could tell by just looking at him. He has been in the Navy for four years and he hasn't done a damn thing. He hasn't made rate. He hasn't gone to school. He hasn't tried to learn. Such a waste. If he is lucky he'll get a job at Jiffy Lube or MacDonald's, or someplace. Someday he may look back and regret wasting the past four years of his life, but for some it never clicks.

Master Chief Quin was able to quickly shift from being a sensitive boss who could get misty in front of her female subordinates to a stern Navy Master Chief—perhaps even a "wolf." She was not going to allow a sailor who was not *living and learning*, or able to *suck it up*, get away without at least one last "free ass chewing." The young women who observed Master Chief Quin's exchange with the non-conforming seaman got a leadership lesson. They were learning the leadership practices that Master Chief Quin learned from the "wolves" who taught her. But Master Chief Quin had adapted the lessons she learned to suit her view of reality and her persona; through her earlier display of emotion she was telling her subordinates that you can be a tough master chief who "chews ass" and still be a mother who can express love for her child.

Through their narratives and stories both the Chief Petty Officers at the Senior Enlisted Academy and on the Ship were constantly negotiating a balance between leadership narratives that prescribed conflicting practices such as *live and learn, take care of your people* and *suck it up*. There was an ongoing process of figuring out how to modify, change or reject practices as the participants attempted to make sense, give sense, cope and adapt to the future. The major challenge that all of the Chief Petty Officers expressed was keeping sailors from getting into trouble and then getting them out of trouble once they were in trouble if they were worth keeping in the organization. Even if they were not worth keeping on the team, the Chief Petty Officers expressed feeling conflicted about giving up on them. While Master Chief Quin seemed disgusted with the young sailor who appeared at her door with visible pegs in his ear piercings, she did counsel him to check in with the reserves so he would not lose credit for his time in the Navy, but she was not holding out a lot of hope for him—she was a "wolf" who was ready to throw him overboard and perhaps she had her reasons for feeling pessimistic about his future prospects—she was trusting her gut—her perception was that he did not belong in the organization.

COMPLEX ADAPTIVE ENTITIES AND THE PRACTICE: LIVE AND LEARN

As was mentioned earlier Complex Adaptive Entities are living systems that self organize and adapt. Humans acquire knowledge by sharing information through story and narratives with their offspring, families, communities and members of other storied spaces. Other animals transfer knowledge to a much greater extent by direct genetic inheritance—instinct—that has evolved over millions of years. Complex Adaptive Entities acquire information through interacting with their environment, identifying regularities and condensing those regularities into ways of behaving in response to the world. Humans share their knowledge through language—narrative interpretation—and because humans have had the ability to share their knowledge through speaking and writing for thousands of years, the learning of rates of humans as a species has been explosive, giving humans dominion over the planet. Through stories and narratives humans can describe how they struggled with a problem and prevailed—these are the stories humans like to tell. The narrative *Sh*t happens* with its attendant practice *live and learn,* illustrated in the previous stories describes a "storied space" that functions very much like a Complex Adaptive System at work.

ATTRACTORS—PATTERNS OF BEHAVIOR THAT CAN LIMIT OR ENHANCE GROWTH

Some of the narratives and practices of the Chiefs included here functioned as "attractors"—that is to say, patterns of behavior that can limit or enhance growth. Some attractors are stable while others are unstable, and there is another type of attractor sometimes called a "strange attractor" that can be both stable and unstable. Strange attractors operate at the edge of chaos. They provide the entity with the opportunity to break with a stagnant pattern and achieve novelty. In business organizations achieving novelty is often referred to as "thinking outside of the box." There was evidence in the stories and narratives of Navy Chiefs that the behaviors the narratives prescribed could be stable even while the narratives and stories that contained them were changing. Many of the stories reinforced a dominant narrative and as new information was received it was adapted to fit the existing narratives and practices. But there was also evidence of unstable patterns and evidence that stories had the ability to function as strange attractors thereby resulting in novel approaches to problem solving that provided the entity with a whole new way of responding to its environment.

Stories such as *New Chief Brings Hope* or *Master Chief Cal's Story*, described how the protagonists in the stories broke free of a stagnant, or destructive, patterns and adopted new patterns that enhanced adaptability. Often these new patterns were established at a time of crisis or chaos for the entity. Master Chief Cal described how he either had to change his behavior or he would not survive in the organization. Senior Chief Dee described a new supervisor who was the catalyst that enabled her to break with a stagnant pattern. The stories *We're Gonna Ace It*, and *Dude What Happened*, described behaviors that were harmful or held the group back, and prescribed behaviors to break from those patterns so that the group could prosper.

As can be seen from the stories and narratives included here, the leadership narrative *Sh*t happens* with its prescribed practice *live and learn* was particularly resilient, enhancing the adaptability of individuals and storied groups. It demonstrated double loop learning—learning that is reciprocal and on-going as the entity accepts the instability that is seen as inevitable and responds with practices that help it thrive.

You gotta keep your head on a swivel. You never know when BAM—all hell is going to break loose

Senior Chief Dave

An often repeated cautionary phrase in the Navy is "keep your head on a swivel." The message is stay alert, remain aware, and anticipate potential problems because the work environment is inherently dangerous and the unexpected will happen. Other phrases such as "Nothing ever happens exactly as planned," "Prepare to be surprised," and "Shit happens," reflected this leadership narrative and practice.

Stories like *USS Kitty Hawk Mishap* and *Jet Blast Deflectors Malfunction*, were accounts of accidents, called "mishaps" that reflected the chaos that follows an unexpected event. Some of these stories were tragic and even terrifying. But other stories that reflected this narrative were told with a humorous tone like *Dippy Sippy Donut Guy*. And a few of the stories that contained this narrative were nostalgic and could even be described as romantic.

In the following epic story 9/11 on the Connie, Senior Chief Tracy, a student at the Senior Enlisted Academy, described how he and his shipmates made sense of the unexpected events that unfolded on September 11, 2001 and the days that followed. In this story he expressed the need to be flexible in this organization. He described how he was willing to make personal sacrifices—*suck it up*—to carry out the mission of the Navy to respond to the terrorist against attacks against the United States.

9/11 ON THE CONNIE

"9/11" was actually the first day I wore khakis [his new Chief's uniform]. I will never forget that day as long as I live. I was stationed on the Connie [the aircraft carrier USS Constellation]. We were transiting from Hawaii to San Diego, coming off Westpac [six month long Western Pacific deployment]. I woke up that morning and put on a brand new pair of starched khakis. I was lookin' good. I walked into QA (Quality Assurance), which is where I worked, and everybody was watching the TV. And I was like, "What movie is this?" And one of the guys is like "This ain't no movie, this is real." And I was like "What are

you talkin' about?" And so throughout the course of the day all the TVs (on the ship) were on and there was news coverage of the planes hitting the towers and hitting the Pentagon and crashing in Pennsylvania. It was a bitter day for us. There we were and there was our country being attacked. We wanted to do something. And I am changing my uniform from blue to khaki, ya know? I was getting pinned [the ceremony at which the newly selected Chief receives the anchor collar devices of a Chief]. But then you know we had all this chaos. Oh, and on top of everything else, we had tigers on board [civilian family members and friends] for the last leg of the deployment from Hawaii to San Diego. So we had all these young kids and civilians on board and they are all watching us to see how we're reacting, like "What happens now?" And then they started asking questions. And I'm a brand new Chief so now all of a sudden just because I am wearing khakis I'm supposed to know something. And the guys in the shop are lookin' at me and asking, "What are we going to do Chief? What's America going to do?" And I am thinking "Damned if I know what the hell we're gonna do." And that's hard. They were expecting me to have some answers. We were supposed to come off cruise, but now the whole mission of the ship would change because of what was happening. The rumors started to fly. Stuff like, we're not pullin' into port, we're not gonna see our families, we're gonna load up and go out and fight. Ya know how it goes? But I'm thinkin', "Hell, we've got all these civilians on board we've got to do something with them." So we pulled into San Diego "darkened ship" [with no lights on]. And that was the first time that an aircraft carrier ever pulled into San Diego "darkened ship." It was eerie, this big ship, pullin' in like a huge shadow. And none of the families knew we were comin' because as soon as something like that happens they of course shut down all coms [communication] with the beach. We unloaded our tigers [civilians] and started to load up our ordnance to go out again. But somehow the word got passed and some of the families started to arrive. So a lot of them came down to the ship. But we only stayed long enough to load our ordnance and then we went out again.

Senior Chief Tracy described the events that took place on September 11, 2001 as "chaotic," they created dissonance in the life of the Senior Chief, he was struggling to make sense and give sense to those around him. Like many Americans on September 11, 2001, Senior Chief Tracy was experiencing a combination of emotions: confusion, pride, fear, anger, frustration and sadness. But perhaps unlike many Americans he had a way to respond to the attack that helped him make sense. He believed he had an important role to play—a job to do that was a direct response to the attack and that helped him cope. Yet he had to make sense of the chaos fast because now that he was a Chief Petty Officer his subordinates were looking to him to help them make sense. The crew was not angry that they would need to go back out to sea without seeing their loved ones. They had been trained that *the unexpected will happen* so they needed to *suck it up* because *they were part of something bigger than themselves*. He implied that, despite the sacrifice entailed, they wanted to do something in response to the attacks. They wanted to fight back.

September 11, 2001, was not a typical day on the ship. It was not a typical day in the lives of Americans. The attack on the United States on September 11, 2001, was an extreme surprise—it was a shock. *9/11 on the Connie* prompted similar personal stories in the rest of the Chief Petty Officers. Through sharing their "9/11" stories the members of the group made sense and gave sense. They affirmed their collective values as a group. They affirmed their belief that the mission of the Navy, and the ships they were on at the time, was courageous and noble. Their stories reflected their principles and practices. They implied that when ordered to do so Navy people must *trust*, following the orders of the officers appointed over them and their Commander-in-Chief (President of the United States), and their families were expected to *suck it up*, and *trust*, when they were sent back out to sea.

Senior Chief Tracy's 9/11 story encompassed events that took place over several days and compressed them into a compact narrative with a beginning, a middle, and an end held together by a plot, sub plots and plot twists. The events themselves did not have a beginning, a middle and an end, until Senior Chief Tracy through retrospection and sense making created a causal connection between them resulting in a story. Through creating connections and making retrospective sense of what occurred, Senior Chief Tracy created and reinforced a view of reality for himself that fit with other frames of reference he had formed over the years. Through sharing his story he helped create, or reinforce, a collective view of reality for the group that enhanced the bonds between members of the group, thus a sort of collective sense making took place.

September 11, 2001 held special significance for Senior Chief Tracy because it was the first day he "wore khakis", that is to say it was his first day as a Chief Petty Officer. When a Petty Officer is promoted to the rank of Chief Petty Officer they change uniforms, from the blue uniform of a Petty

Officer to the khaki uniform of a Chief, a highly visible symbol of their change in status. Along with the change of uniform comes increased authority and responsibility in the organization. Senior Chief Tracy had to make sense of the chaos fast because now others were looking to him for guidance and sense making.

The symbolism of changing from "blues" to "khakis" is huge in the culture of this organization. Senior Chief Tracy explained that one day you are "one of the guys" and the next day "you're expected to know something—people are looking to you for the answers." Promotion from within the ranks to a management position is awkward for the person making the transition in most organizations, but it is especially tricky in a hierarchical organization like the Navy where everyone's rank is immediately visible. No doubt one's position in the hierarchy also determines who the storytellers are and who their audiences will be. Does rank also determine who will have the privilege of defining the organizational view of reality and which stories are told as reflective of the truth? Most likely those in positions of authority do influence the "grande narratives" (Boje, 2001) that are told within the organization. Such stories can help the organization make sense but they can also prevent the organization from seeing divergent views of reality that could result in novel solutions to problems.

Most Americans, who are old enough to remember, vividly recall what they were doing on September 11, 2001. The terrorist attacks that took place on that day are "trigger" events for most Americans that result in what Boyce (1996) calls "touchstone" stories. The events that happened that day "trigger" memories and as a result they can produce a stories that resonate with others who also experienced the chaos of that day. September 11 stories create connections between the attacks that took place and people who were separated by hundreds or thousands of miles that day. Most people cannot remember what they were doing on a specific day last month, but years later they can still vividly recall September 11, 2001 and in telling their personal "9/11" stories they are transported back to that day. Their stories help them make sense of the dissonance—the chaos created in their lives—and communicate that sense making to others. Their personal "9/11" stories help them clarify their principles and thus prescribe maps for how to behave in the future.

Other trigger events for Americans old enough to remember them include: the attack on Pearl Harbor, the assassination of President Kennedy and the assassination of Martin Luther King Junior. Natural disasters such as earthquakes, hurricanes or tornadoes are trigger events that result in stories that help people make sense of chaos and coalesce around a narrative surrounding those events that define their view of what happened—their collective view of reality. Organizations have their own trigger events that result in touchstone stories that help the members of the organization make sense, cope and move forward; they create a bond and a connection within the group. If the same emotions are shared in the stories, such as shock, confusion, fear, sadness or anger they create a bond. There are also trigger events and touchstone stories that are unique to organization subcultures. Stories surrounding such events create a bond between individuals and groups of people who are part of the same culture or subculture. Of course, the events that took place on September 11, 2001, while still "trigger" events would create entirely different "touchstone" stories for the members of terrorist group in Pakistan, but they would serve the same function helping the members of the group make sense, give sense, cope and navigate into the future. And when an individual's stories are not compatible with the collective narrative created by touchstone stories the individual will very quickly realize that they are out of step with the others in that particular storied space.

Most groups as small as families or as large as countries share trigger events that create touchstone narratives that in turn define the group's view of reality and reflect a complex panoply of commonly held values and principles. Sometimes all it takes is a word or a phrase to evoke the memory of what happened, and the entire story of what happened and its meaning is contained in that word or phrase. Touchstone stories can thus help people make sense of unfolding events and inspire action. Individuals living with the dissonance of seemingly incompatible narratives may not be able to cope and may strike out at the group, or even become self destructive, but recognizing the dissonance and understanding it for what it is may help humans cope with it and emerge stronger.

9/11 on the Connie triggered other September 11 stories in the group. Indeed, each member of the group had their own "9/11" story to share. Following is another "9/11" story. Although this story was sparked by the same events it is a different kind of story that served some different functions on the day it was told.

9/11 with the Marines in Johnston was told by Senior Chief Dave, whose specialty is electronics and internal communications. On September 11, 2001, Senior Chief Dave was on temporary duty with a Marine Corps unit in Johnston, Pennsylvania, where one of the planes headed for the Pentagon crashed. His duties involved setting up a communications network and training the Marine technicians in the unit to maintain and repair it. All Marines are trained, skilled weapons handlers, in contrast most sailors have technical specialties other than weapons handling. In the following story the contrast between two military sub-cultures—the Marine Corps and the Navy—contributes to the chaos and confusion in Senior Chief Dave's personal "9/11" story.

9/11 WITH THE MARINES IN JOHNSTON

I was working for an admiral's staff at the time, in IT [Information Technology]. I was a fiber optic technician and we were out in Johnstown, on a base southeast of where the plane went down. And that day we were actually out at the regional airport because the Marines had just moved there and they needed to be set up so that they could run operations. I'm not trained to carry a weapon. I am trained to lay fiber optics. And we got word over the loud speaker to get our weapons and get out. Apparently the chain of command got the word that there was a plane and it may be coming our way and the planes had already hit the other targets so nobody knew what was coming next. Anyway, they told us to grab our weapons and take cover in the woods. It was absolute chaos. It was like a bomb already went off. People were runnin' all over the place. I didn't know how to fire an M16. I had never even held an M16 before. And all of a sudden this Marine hands me an M16 and tells me to take cover. And I said, "What am supposed to do with this?" And the guy who gave it to me gives me this (disgusted) look and says "What do you mean what are you supposed to do with it, you shoot it, now take the damn thing and git!" I wasn't gonna argue with him, so I took it, and latched onto the nearest Marine and stuck to him like glue. We ran into the woods. It was a crazy. I transferred shortly after that and at my next command there was a range (shooting range) and I had never qualified on a weapon before, but after that incident I went out to the range and qualified (learned to shoot) because I figured I'd better just in case something like that happened again.

9/11 with the Marines in Johnston is an epic story with a lighter tone than *9/11 on the Connie.* Senior Chief Dave described how he muddled through the chaos of the day and survived. His story not only included the narrative *the unexpected will happen* and its attendant practice *head on a swivel*. He did not need to repeat the narratives *you are part of something bigger than yourself* with its attendant practices *suck it up* and *trust*, the group had already coalesced around those points in several of the other "9/11" stories. *9/11 with the Marines in Johnston* lightened the somber mood that prevailed during many of the other "9/11" stories. It served as a natural transition to another topic at a time when the "9/11" stories had served their purpose and the group was ready to move on to another topic.

"SEMPER GUMBY" (STAY FLEXIBLE)

One of the Chiefs coined the somewhat sardonic phrase "Semper Gumby" to communicate that you always need to be flexible in the military because the best laid plans are subject to change. This aphorism—a take off on official military mottos that begin with the Latin "semper" (always) such as the Marine Corps motto "Semper Fidelis" (always faithful) combined with reference to "Gumby" the goofy cartoon character and rubber doll whose strength is in his flexibility—seemed to strike a cord with the other Chiefs. Many of the stories that reflected the leadership practices *head on a swivel* and *stay flexible,* were about how the protagonists, or other characters in the stories, made sense of, and coped with, surprise, change and the unexpected. They communicated the need to be able to react quickly to rapidly changing situations in a sometimes chaotic environment. But many of the stories also communicated that some change, while inevitable, did not happen quickly in the lumbering bureaucratic environment of the Navy. The stories communicated that big policy changes happened over time, giving members of the group time to coalesce around a new narrative and adjust. The stories also communicated that those who could not adjust their personal narratives to the changing narrative of the organizations would need to adapt or move on. In the following story, Senior Chief Dee, a student at the Senior Enlisted Academy who also told the story *New Chief Brings Hope* described a beloved "crusty old Master Chief" who was struggling to adapt to a changing Navy.

MASTER CHIEF PT'S IN KHAKIS

Master Chief Kelly was a crusty old Master Chief, but in many ways he was not like the other Chiefs. In those days Chiefs were out of sight out of mind—they let the LPO [Leading Petty Officer] do everything. But Master Chief Kelly was not like that. He was always doing stuff for us, and takin' care of us. It was hard to get a car in those days so he was always driving us where we needed to go. He would organize division parties and pick everyone up and take everyone home. He would drive you to the airport if you needed a ride. Like a lot of Chiefs in those days he smoked like a chimney and he was always holding a coffee cup that looked like it hadn't been washed in years. And if someone tried to wash it there was hell to pay—he said the crust gave the coffee better flavor. So one day we were scheduled to do our PT test [semi-annual physical training test] and we were all out by the gym in our PT gear and the Master Chief shows up in his khakis and work boots and we were like "Master Chief where are your sneakers, where is your PT gear?" And he was like, "The Navy didn't issue me no god damn PT gear in my sea bag." And so he did the run and the push-ups and sit-ups and the whole test in his khakis and work boots. He was huffin' and puffin' and beet red when he finished but he passed it. And as soon as he finished he pulled out a pack of Camels and lit up.

The beloved "crusty old Master Chief" was passive aggressively complying with the changing Navy policy with respect to fitness standards. He said the Navy "didn't issue me no goddamn sneakers in my sea bag," justifying his behavior. But the new fitness standards would be phased in over time and "crusty old Master Chiefs" who could not be "Semper Gumby" would retire from this young organization where the average age is 20 and you are an "Old Yoda" when you are in your 40's. The "Old Yoda's" pass on the narratives that have evolved over time and they also prevent some change from happening too quickly—therefore, big change happens gradually.

Both Senior Chief Dee and Master Chief Quin described a Navy that had changed its policies and attitudes toward women since they were young sailors, but this change happened over time as well, allowing the organization to adjust. Both women said that they disagreed with the characterization of the Navy as "a man's world," even though the Navy is twenty percent female. Senior Chief Dee said,

The guys who thought a ship was no place for a woman are gone. They were dinosaurs. That kind of attitude is not tolerated today and the Navy can do that—zero tolerance—if you act sexist or racist you gotta change or leave. You might not be able to change the attitudes of some people but you can change their behavior by just not tolerating it.

One day, while crossing the Hangar Bay of the Ship, Master Chief Quin observed and commented on a petite female seaman in coveralls driving a small tractor with a huge airplane called a "rhino" in tow. As the young seaman adeptly maneuvered the "rhino" into an incredibly tight parking space, I commented that I never thought I would see women doing this type of work on an aircraft carrier. Master Chief Quin replied, "Yeah. I know what you mean, but you've got to remember that this is the only world these kids will know—they will not have memories of the time when women did not do these things."

Senior Chief Pay, a Boatswain Mate in charge of the Deck Department, reflected upon how the Navy had changed since he joined,

When I was a young sailor if you told the Chief that you needed time off because your wife or kid had a problem, the typical response was "If the Navy wanted you to have a wife they would have issued you one in your goddamn sea bag!" But those days are gone. There was still a lot of racist shit going on when I joined. Somebody would call someone a name and there would be a fight that would turn into a small riot. And of course there were no women on ships. Now I've got more women in the Deck Department than men and there are more married sailors than single sailors—it's hard to believe. But, ya know, it works, it's got to work. It's not like you don't have any problems with female sailors or married sailors—they are just different than the problems you have with single male sailors. The guys go out and get drunk and don't show up for work, or they get into fights and after a couple of incidents they get kicked out so you got to deal with that. The women have other problems—like they turn up pregnant and then we lose them—that's a problem. And the married sailors have family issues. We always seem to be losing people for one reason or another. It's a problem but you just gotta deal with it.

The stories that reflected the need to accept chaos, change and surprise, and the need to adjust behavior accordingly, also communicated the message that if you cannot adapt to change you will need to leave the organization. The prescribed practices of *head on a swivel, stay flexible, live and learn, suck it up* and *trust* enabled members of the group to adapt to the chaos and change that was viewed as inevitable.

CHAPTER SUMMARY

There are times of crisis and chaos in any organization when *Sh*t happens.* Members of a storied space who are *flexible* are best able to cope with crisis and chaos. Since crisis and chaos are inevitable the practice of *head on a swivel* reflected the need to remain alert at all times. Finally, tolerance for mistakes with the belief that the individual and the organization can learn from mistakes—and grow—was reflected by the practice *live and learn.*

The narrative *Sh*t happens* and its attendant practices, in addition to the narratives and practices discussed in the rest of Part III work together to provide behavioral maps for the Navy Chiefs whose stories are examined here.

LEADERSHIP REFLECTION

There are times of crisis and chaos in all organizations when "Sh*t happens." Lessons from the study of natural and living systems indicate that learning from mistakes, staying alert and remaining flexible can result in double loop learning and novel adaptations that propel growth.

1. *What stories circulate in your organization about the leadership's reaction to mistakes? What is the over arching narrative with respect to making mistakes? Are there "live and learn" stories in your organization? How can you use these stories to instruct or inspire as your organization tackles current and future challenges?*

2. *What are your personal "live and learn" stories? How might those stories be instructive for others?*

3. *What stories do you have in your repertoire about adapting to good, or bad, leadership? How have these stories informed your leadership style? How might these stories be instructive to others?*

4. *Are there stories and narratives in your organization that are functioning like "limiting attractors"—patterns that limit growth? Are there stories and narratives that could function as "strange attractors"—and inspire people to break with stagnant patterns and think outside of the box?*

5. *Are you, as a leader, limited by stories and narratives you have come to accept as reality? What stories and narratives are preventing you from breaking with stagnant patterns and finding novel solutions to problems? How might the stories of others help you imagine new realities and novel solutions to problems?*

6. *What are the stories about crisis and chaos in your organization? How have you or the organization handled crisis and chaos? What do these stories say about how you, or your organization, will handle crisis and chaos in the future? Do they reflect how you want to handle crisis and chaos in the future?*

7. *What are the trigger events that have resulted in touchstone stories in your organization? Are there different versions of these stories? Is there a need to make sense of events that have effected your organization? If so, how can you use narrative and story to make sense and give sense to others? Are there touchstone stories that can enhance the bonds between the members of your organization?*

8. *Living systems are never static—change is inevitable. What are the change stories in your organization? How can examples of how your organization has adapted to change in the past help the organization cope with change in the present and prepare for future change?*

9. *Once behavior changes, attitudes will follow. But attitudes will not change overnight. What behavioral changes need to take place in your organization to pave the way for attitude changes? What stories can you use to facilitate change in attitude?*

10. *What are the underlying themes and dominant narratives in the stories you identified above? Dominant narratives that may be at play: "live and learn," "stay flexible," "we've always done it this way," "don't rock the boat."*

(See Appendix D (Storytelling guidelines) to hone storytelling skills)

Chapter 9

DO NOT TAKE YOURSELF TOO SERIOUSLY, BUT DO TAKE YOUR WORK SERIOUSLY

You get to have all this fun and a paycheck too! I ain't never leaving this organization

Senior Chief Shane

The leadership narrative and practice *Do not take yourself too seriously but do take your work seriously* was expressed by other aphorisms such as "work hard, play hard," "If it ain't fun you're doing something wrong," and "know your craft."

Do not take yourself too seriously but do take your work seriously reflected the principle that maintaining a sense of humor is an essential coping mechanism in this organization. As was discussed earlier, much of the work in the Navy is not glamorous; it can be tedious and grueling. But there was the sense among the Navy Chiefs that while the work might at times "suck," at the end of the day it should be satisfying and everyone ought to be able to take pride in a job well done. Taking pride in work that sometimes "sucks" entails having a sense of humor. Navy Chiefs whose stories are explored here all displayed a sense of humor.

As was discussed in Chapter 6, the type of humor favored by Navy Chiefs was self-deprecating and physical. They also favored "gallows humor" that was often irreverent and sarcastic. But a universal theme that ran through the stories of the Chiefs was that by the time you have reached this level in the organization you ought to be mature and self aware enough to laugh at yourself as well as the organization.

Daniel Goleman in his research and writing on Emotional Intelligence (EI) found that excellent leaders display an excellent sense of humor, "Research on humor at work reveals that a well timed joke or playful laughter can stimulate creativity, open lines of communication, enhance a sense of connection and trust and of course make work more fun," (Goleman, 1995, p. 14). Senior Chief Petty Officers and Master Chief Petty Officers were especially good storytellers and quick with humorous quips that would release tension in a group. The humorous stories were energizing and often cathartic.

Many of the humorous stories were simply entertaining. The Chiefs at times tried to out do each other with humorous stories. Many of the stories were like inside jokes, they were culture specific and would not be humorous to someone who did not understand the Navy, but within the culture of the Navy they were hilarious. The stories that were inside jokes created a bond within the group, again reflecting shared values and membership in the "storied space" of the Chiefs' Mess.

The following story, like *AB or Wanna Be,* and *French Fries,* reflects the type of self-deprecating humor that characterizes Navy Chiefs. It was told by Senior Chief Shane—an engineering specialist who was a student at the Senior Enlisted Academy. This story energized the group and prompted other similar self-deprecating stories.

ALL THIS FUN AND A PAY CHECK TOO

I was raised by my grandparents. My Grampa would tell us his stories about World War II and the adventures he had and it sounded like a lot a fun to me. He never talked about how much he got paid. He just talked about the places he had seen and what he did. We never had much money as kids—being raised by our grandparents and all. We lived in the country. We didn't need much money and somehow we scraped by and didn't know anything different. As soon as I was old enough, I decided to join the Navy because of all the fun my Grampa had. I went off to Boot Camp [basic training]—my first airplane ride—and right off the bat they are giving us stuff. They march us into the Gear Issue and give us all this stuff, skivvies [underwear] and boots and soap and shaving gear, and the like. And we go marching back to the barracks with all this stuff. Well for me it was like Christmas. I never had so much never-been-used stuff before. And they are feeding us three squares a day and the chow ain't bad. I got into some trouble for being a knucklehead once in a while, but boot camp wasn't bad, and I figured I just needed to get through it so I could have some adventures like my Grampa. Well one day they line us all up, call out our names and hand us a check. Well I had never had a check before. I held it and looked at it with my name on it and all and I quickly stuck it in my pocket. And as soon as I got back to the barracks I go to my locker and I hid it in the folds of my clean t-shirts. I was sure they would ask for that check back someday for all of the free stuff they gave us. And a few more weeks go by and they give us another check and this one is for $700 and I have never seen $700 before and I was scared to death so I go and hide it with the other check. And a few more weeks go by and we get our assignments and we get another check and now I am just feakin' out, so finally I turn to this buddy a mine and I says

"Hey man, is this my check, do I get to keep this?" and he gives me a (disgusted) look and says "Yeah, course you get to keep it, you get paid here man!" And I was like "Yahoo! I ain't never leaving. I am stayin' [in the Navy] forever." I could not believe it. So that night we went out and I went wild—we went bowling and I paid for everybody.

Senior Chief Shane had come a long way since he received his first paycheck. He was no longer a naïve country boy. He was an accomplished technician and respected leader. Others in the group could relate to his story, they nodded and chuckled. After the telling of several such stories the group discussed how this type of story communicates to others that it is okay to laugh at yourself and that laughing at yourself can help others feel comfortable doing the same. Self-deprecating stories demonstrate a willingness on the part of the storyteller to be open and even vulnerable within the group; as a result they communicate trust and strengthen the bond of the group. And when humorous stories increase trust they help to foster a playful environment that is conducive to creative problem solving.

In the following story, Senior Master Sergeant George—an Air Force exchange student at the Senior Enlisted Academy—described an incident that took place when he was a load master—the person in charge of the people and cargo on an air mission who also monitors safety and assists those parachuting out of the aircraft. In this story he was training others to be load masters while participating in a "drop" of Navy Seals—a highly trained elite force of Navy frogmen who parachute from aircraft into water to carry out special forces missions. Senior Master Sergeant George was not Navy but his story is included here because it suggests that there may be commonality across service lines at this level of senior enlisted leadership.

LOAD MASTER TEACHES SITUATIONAL AWARENESS

I have a story that I use with my students about situational awareness. I use this story to teach situational awareness. I tell them you can lose your sense of situational awareness no matter what level you're at whether you are a junior airman or a Master Sergeant. And I tell them about the time that I was on an aircraft drop for some Navy Seals and I lost my situational awareness. When you are up that high you need to have a restraining harness on the load master and you need to hook it to a tie down ring so that when you open the doors you don't inadvertently fall out of the aircraft. I had three students in each door and a couple of other instructors I was training and of course all these Navy Seals. And at the last minute one of my students tells me he left a piece of gear we needed on the ramp [the hinged ramp of the aircraft that opens for the drop] and asks should he go get it. And I tell him "Sit tight, I'll git it." So I take off running toward the ramp because I only have a couple minutes before we need to start the drop. Well, I'm still hooked up. And everybody's lookin' at me. I am the guy in charge and I am supposed to know what I'm doin'. I get up some good speed and I get about 18 feet, I'm on the ramp, and WHAM, I go flyin' back through the air head first like one a those paddle balls on a rubber band, cause I'm still hooked up and I SLAM down on the deck on my head. Lucky I had my helmet on. I got knocked out for a couple a seconds. I came to, shook my head, and I look at my students and their eyes are poppin', they are all just stunned, and I look at the Navy Seals and they are all laughin', I mean they are crackin' up, and one of the Seals says "That was the funniest thing I ever did see, could you do it again?" And then everyone starts snickerin'. It took me a couple a minutes to get my composure after that.

Senior Master Sargeant George used self-deprecating humor in this story but he didn't use it merely to entertain, energize or enhance the bonds within the group, he explained that he uses this story to train his students. He even provided the moral and theme for the story in the introduction—"this is a story about situational awareness." This story is like the "don't be the guy…" stories discussed in Chapter 5. Senior Master Sergeant George said he tells students this story so that hopefully they will not make a similar mistake. *Load Master Teaches Situational Awareness* was a self-deprecating "slap stick" story. If Senior Master Sergeant George had broken his neck this would not be a funny story—but framed as it was with surprise, colorful characters and danger, it was very effective as a humorous story with a redemptive message. Senior Master Sergeant George played the fool in this story but he redeemed himself by using his mistake to illustrate a point and teach a lesson.

While the practice *Do not take yourself too seriously* was reflected in humorous stories that energized the group and enhanced a playful atmosphere, when combined with the caveat *but do take your work seriously* it sometimes presented a duality. What does "work hard, play hard" really mean? How does the individual or the group know when the play has gone too far, or when it is time to get back to work. In some of the stories "playing hard" was used to justify risky or even harmful behavior. Some humorous stories described the outrageous escapades of sailors on liberty that were not in the best interest of the individual or the group. The Chiefs were at times clearly conflicted about what they perceived as a duality between working hard and playing hard, when playing hard for young sailors resulted in risky behavior that could result in physical injury or disciplinary action.

Chief Brent, an Aviation Boatswain Mate, told the following humorous epic that could have resulted in disaster, and would likely result in disciplinary action if it happened today. Chief Brent is stocky and tall—he towered over most of the sailors he supervised—but he had a boyish demeanor. He told this story over a cup of coffee in his office after touring his area of responsibility on the ship to supervise work in progress. Several sailors working nearby clearly seemed delighted that Chief Brent was in his storytelling mode.

BIG PAPA IN WAIKIKI

I was stationed at a H-46 squadron [helicopter squadron]. We had just returned from deployment and we had done a particularly good job. Me and my LPO [Leading Petty Officer] had kicked ass and it hadn't gone unnoticed. A couple of the Lieutenants in the squadron told us that they wanted to show their appreciation so they took us out to Duke's in Waikiki. And they told us that the drinks were on them and we could order anything on the menu. So we proceeded to order everything on the menu—I mean everything. We ordered and proceeded to drink every single drink on the menu, from mai tai's to pina coladas, they kept buyin' em and we kept drinkin' em. I evidently made it through the entire drink menu. When I woke up I was layin' in the sand on the beach with my stomach distended like a beach ball. My LPO, who was Jamaican, was sittin' there singin' and playin' the ukelele and there were these two little Hawaiian kids singin' and slappin' my stomach like bongos.

Chief Brent had quite a repertoire of outrageous drunken sailor stories and evidently had quite a reputation for telling such stories. A couple of the Chiefs on the Ship seemed a bit uncomfortable that Chief Brent was participating in a research study on story, probably because they thought he would tell stories that in their opinion "should not be told."

But Chief Brent was not a one dimensional character. While he seemed to like to entertain his audience with tales of his outrageous escapades his behavior in many ways contrasted with his stories. In the discussions that followed Chief Brent's tales about his antics as a young sailor—some of which could not be repeated here—he shared his version of a *live and learn* practice that appeared to have influenced the way he lived and worked. At one point in our conversation he pulled a photo out of his desk drawer of himself as a handsome young sailor and the home that he had purchased. He explained that after "pissing away five years (as a young sailor) and having nothing to show for it," he decided "it wasn't so cool to get wasted all the time," so he started a savings allotment; as a result he was able to buy a nice home and make several other wise investments that paid off over the years. Chief Brent was proud of his accomplishments. No doubt he has shown others the photos in his desk drawer and shared the lessons he learned after "pissing away five years."

It is likely that Chief Brent was telling me the type of macho story that he thought I expected, that is to say stereotypical "sea stories" about the outrageous behavior of sailors on liberty that are told for self-aggrandizement and to amuse others. Chief Brent certainly had plenty of those stories in his repertoire. But it was also obvious from the larger narrative that surrounded his stories that Chief Brent had adopted a complex blend of practices to cope. His behavior at work probably contradicted many of the antics he described in his stories and that contradiction was no doubt apparent to those mature and insightful enough to see the contradictions. The messages that Chief Brent's subordinates take away from his stories are unknown. No doubt the maturity level of the audience largely determines the take away.

As the stories and behavior of Chief Brent illustrate humans are not one-dimensional, they are complex creatures. A unique combination of stories and narratives determines an individual's identity. While there are similarities between and among individuals and a natural desire for individuals to coalesce as a group, all groups are composed of individuals with different needs, desires and motivations. Therefore it is important to understand the narratives guiding the individuals as well as the group in order to tap into the full potential of the storied space.

CHAPTER SUMMARY

Many of the humorous stories collected for this book reflected the leadership narrative and practice *Do not take yourself too seriously, but do take your work seriously*. Most of the humorous stories were self-deprecating and many were physical—slap stick—or sarcastic "gallows humor." Many of the humorous stories were cathartic and energizing; but many also contained valuable lessons. They helped the tellers and their audiences make sense, especially those that contained the practice *live and learn*. They helped the tellers and audiences laugh at mistakes and see how those mistakes resulted in growth. Many of the stories added levity to the workplace environment and helped people cope in the present—they were fun and entertaining. Humorous stories energized and strengthened the bond of the group—especially those that were self-deprecating—because they communicated trust and a willingness on the part of the teller to be open or vulnerable. Many of the humorous stories that were told were like inside jokes; that is to say they would not be funny to those outside of the Navy thereby affirming that the teller and the audience were part of the same storied space. The lessons learned from the humorous stories of Navy Chiefs is that having a good sense of humor is essential in this organization.

LEADERSHIP REFLECTION

As was discussed in Chapter 4, humor is tricky. One person's joke may be another person's insult. Humor can be passive aggressive and used to marginalize individuals or groups of individuals. But used appropriately humor can energize the group, inspire trust and enhance the playful atmosphere of the work place resulting in creative ways to accomplish what needs to get done. The ability to admit to personal mistakes, learn from them and laugh at them is the sign of a mature leader. A leader's appropriate use of self-deprecating humor can be especially effective, strengthening bonds within the group and encouraging others to try novel approaches without the fear of reprisal.

1. *What humorous stories circulate in your organization? What values do they reflect? Are they self-deprecating? Are they physical? Are they ironic or satirical? How could you use these stories to energize the organization, enhance teamwork and foster a creative, playful atmosphere?*

2. *Listen carefully to the stories that people laugh at in your organization, are they ever used to marginalize individuals or groups outside or inside of the organization? How do these stories hurt or help the organization? Do people laugh at stories that reflect passive aggressive behavior? What messages do these stories contain about changes that perhaps need to be made?*

3. *What is your brand of humor as a leader? Are there stories about mistakes the organization has made that in retrospect are really rather humorous? Do you*

as a leader help the organization laugh at itself? Do you as a leader foster a playful environment in which members of the group feel comfortable sharing a laugh?

4. What stories in your organization reflect members' attitudes towards work? Do members of the organization take their work seriously? Do stories circulate that reflect pride in a job well done? Are there stories about high standards of excellence and quality performance?

5. What are the themes that connect the stories you've identified and what are the dominant narratives? Dominant narratives that may be at play: "A job worth doing is a job worth doing well," "Good enough for government work," "Work hard, play hard," or "All work and no play."

(See Appendix D (Storytelling guidelines) to hone storytelling skills)

Chapter 10

SOME STORIES SHOULD NOT BE TOLD

Some of these guys tell stories about things that they did back in the day, and I don't even know if half that shit is true, but it sends the wrong message.

Chief Blane.

There was one additional leadership narrative—*Some stories should not be told*—that was observed, and alluded to, through comments and reactions both on the Ship and at the Senior Enlisted Academy. This leadership narrative was not frequently repeated in the actual stories told; rather it was an unspoken sentiment that was present in attitudes, demeanor and dialogue with some of the Chiefs. It is discussed here as a leadership narrative because although it was not frequently articulated in the stories the

Chief Petty Officers told, it did occur repeatedly in nuanced ways, and was openly articulated by some Chief Petty Officers.

There is a saying in the Chiefs' Mess that "What happens in the Chief's Mess, stays in the Chiefs' Mess." One Master Chief put it this way, "There is a pact among us that we can disagree within the mess. And believe me it can get bloody in there. But those discussions and disagreements do not leave the Mess." The sacrosanct confidentiality of discourse that takes place within "the Mess" was perhaps an overarching narrative that influenced the wary response to my interest in the stories of Chiefs and how they function within the Navy; and that wariness was understandable. Chiefs sometimes do use unorthodox methods to get the job done, and they do not want the world to know about those unorthodox methods because if they are exposed they will be told to stop what they are doing and follow official, established procedures.

As was addressed in Chapter 2, Chiefs operate in a network that is accepted, and even encouraged, but not officially sanctioned. Many organizations have shadow organizations that are involved in getting work done but are not recognized on official organization charts. In the Navy there is a symbiotic relationship between the tight structure of the official command and control bureaucracy and the Chiefs' Mess that networks around the bureaucracy to get what it needs to accomplish the mission—both structures serve their purpose. Chiefs often operate in a creative gray area where it is "better to beg forgiveness than ask permission."

But beyond the desire to protect the sacrosanct nature of dealings within the Chiefs' Mess there was an expressed desire to control story and the behavior it might reflect or inspire. One of the Chief Petty Officers on the Ship (pseudonym: Chief Blane) openly addressed his contention that "some stories should not be told." Chief Blane did not volunteer for an interview, or agree to be observed on the job, but he approached me to voice his concerns one day as I was leaving the Chiefs' Mess after lunch with the following comment,

I just think that some of these so-called sea stories should not be told. They are dangerous. They send the wrong message. Some of these guys [Chiefs] tell stories about things that they did back in the day, in places like Olongapo [Philippines] or Mombasa [Kenya], and I don't even know if half that shit is true but it sends the wrong message. These kids hear these stories and they go down to TJ [Tijuana, Mexico] and try to do the same thing and they get into trouble.

Another female Chief Petty Officer at the Senior Enlisted Academy hinted at a similar concern. Although roughly ten percent of the Chief Petty Officers at the Senior Enlisted Academy were female, initially there were no female volunteers for the group interview sessions on story and narrative. I was puzzled by why women were not volunteering for the group interviews, so on a coffee break I approached one of the female Chief Petty Officers who was a particularly vocal participant in classroom discussions, and asked her why she did not seem to want to participate in a group, or individual, interview. She replied with a nervous laugh, "Who me? I don't have any stories." Then she rolled her eyes, pointed to a group of male Chief Petty Officers engaged in animated conversation across the room, and said, "Those guys are the ones with the stories, not me." Another female Senior Chief at the Senior Enlisted Academy, a staff member, reacted the same way when I approached her in the staff lounge. She laughed nervously and said, "I don't have any stories." To which I replied, "Come on Senior Chief, you've been in the Navy longer than I was and I have stories, I know you have stories." She just smiled, demurred, and changed the subject.

Were the stories of the two female Chiefs too personal to tell someone they perceived as a stranger—an outsider? Were they protecting the confidentiality of the Mess? Were they more intimidated than the male Chiefs by a retired female Navy commander? Have they come to terms with what they have had to "suck up" over the years to succeed as a woman in a largely male environment? Were they afraid of being exposed or vulnerable? Have they learned to be wary of the open display of emotion that might come with sharing a story? Are the leadership principles and practices of female Chiefs different from those of their male counterparts? As a sub group of the Chiefs' Mess did they decide that they did not want to participate? Did they think research about story and narrative was a frivolous waste of their time? There is no way that I can answer those questions, but as the researcher who conducted the Navy's first study of sexual harassment in 1979, I heard plenty of shocking stories from women who were trying to cope in a predominantly male environment. Indeed, the organization has come a long way since those days, but a female Chief Petty Officer today—an "old Yoda"—likely has endured a great deal, "sucked up" a lot and found ways of coping which may have entailed being wary of whom she shares stories with.

In fairness to female Chiefs it is important to note that the mostly male hierarchy of the organizations involved was initially wary of participating in the research discussed here. Initial letters of inquiry went unanswered but once they were followed up with phone calls and face-to-face meetings with the heads of the organizations the response was positive. Participation at both the Senior Enlisted Academy and on the Ship was slow initially but once key individuals saw potential value in the research participation increased.

There were likely a number of reasons for the wary reactions described above, including protection by the brotherhood of the confidentiality of the Mess. While I was an organizational insider in some respects I was not a part of the storied space of the Chiefs' Mess and never had been. I needed to earn the trust of the participants before they would engage with me.

Also what I was doing was probably poorly understood, or not sufficiently understood, and that was due to my inability to effectively communicate my purpose to those who were put off by the word "story." To many in the Navy "story" brings to mind "sea story" and sea stories are synonymous with "tall tales" that sailors share to bolster their egos. Many members of the Navy approached for the study had a hard time accepting the notion that story analysis could yield any legitimate insights. But many of those who participated in the study, or gave their permission for the research, were able to stretch their understanding of the word "story" to see the potential value of examining story, and they became intrigued. Some immediately grasped the purpose and value of the study and others came around to the idea after discussing it. Once the individuals who were centers of influence, or key nodes in a communication web, assessed the study as low risk and potentially beneficial, many others came around to the idea; but some remained skeptical.

The leadership narrative *Some stories should not be told*, also reflected discomfort with risk and lack of control. This is a typical human reaction. In every group there are risk takers and there are those who are more risk averse and cautious. Both types serve a valuable function in groups as long as the organization finds a balance between both attitudes. But sometimes, perhaps especially so in command and control type organizations, there is an illusion of control especially with respect to communications. Attempting to squelch rumors or prohibit members of an organization from telling stories is futile. Humans are not machines—they do not have an on/off switch. They will always find a way to tell stories. Likewise, spinning a "grande narrative" (Boje, 2001: 10) will not guarantee its acceptance as truth throughout the organization.

Perceived lack of control results in discomfort. Some members of the storied spaces at both the Senior Enlisted Academy and on the Ship were obviously uncomfortable with their inability to control stories. Master Chief Cal and Senior Chief Tray referred to "controlling behavior" and how it was easier to "control sailors' behavior" when the Ship was at sea. Master Chief Cal expressed his frustration that the behavior of the young sailors was beyond his control. The Chief Petty Officers who said, or implied that "Some stories should not be told," were uncomfortable with their lack of control over the behavior that stories might inspire or the negative image of the organization that the stories might project. Or perhaps they were internally wrestling with their own stories and were not ready to reveal them to others.

Stacey (1996), in his application of Complexity Science principles to organization studies, explored how difficult it is for organizations to truly control the actions of their members. He distinguished between behavior that is controlled and people who are in control.

For people to be in control they must be able to specify desired outcomes and identify actions that are likely to produce those outcomes, and then be able to employ negative feedback to keep actual outcomes close to desired ones. People can therefore only be in control in rather limited circumstances (Stacey, 1996: 286).

The Navy Chiefs who are expected to influence the behavior of young sailors, were struggling with their inability to control their subordinates' behavior especially when the sailors were not contained on the Ship. The Chiefs who were concerned about the image that some stories might project could not control the stories or the images—not in a "storied space" that is part of a culture that values free speech.

That being said, there are stories in organizations that should not be told, or at least nipped in the bud. There are stories that should be confidential. There are stories that are counter productive and destructive. There are stories that are demeaning, or degrading. There are stories that are hurtful, or hateful. There are stories that grossly misrepresent the organization. But simply ignoring such stories will not make them go away. It is incumbent upon the leadership of an organization to be aware of the stories that are circulating and remain open to how such stories reflect the undercurrents, attitudes, values and concerns that exist. Those undercurrents need to be addressed or openly confronted when they present a potential threat to the organization. If conflicting stories and divergent views of reality persist and the leadership perceives the dissonance as harmful, it is incumbent upon the leadership to examine why the dissonance persists.

It may be impossible to completely control the stories and narratives that travel through an organization, but that does not mean that the leadership should not confront rumors and gossip when they pose a threat to the organization, or counter such narratives with a more authentic narrative. And timing is crucial—potentially destructive rumors need to be countered swiftly before they cause long-term damage. But many rumors are so far fetched that they will dissipate on their own.

How do leaders differentiate between stories that need to be addressed and stories that will dissipate on their own? Leaders who have developed the good judgment that comes with being aware of the stories that have informed their values will know which potentially harmful stories they should confront and which stories will dissipate on their own.

CHAPTER SUMMARY

Leaders should be aware of the stories that are flowing through their organization and beyond to the environment surrounding the organization. Rumors that do not dissipate may be symptomatic of organization ills that need to be addressed. In healthy organizations observable actions—and the stories they inspire—will support the values espoused by the leadership. When the leadership believes that "scuttlebutt" or rumors are out of line they should examine them for alternate views of reality and be open to engaging in a discourse to dispel them, or counter them with the truth.

Leaders can use their own stories and narratives, and the stories and narratives of others within the organization, to foster behavior and attitudes that will enhance the organization's ability to thrive and prosper. But stories and narratives launched by the leadership need to be authentic or they will be exposed as false. And once falsehoods are exposed, the members of the organization will feel betrayed, and it will be difficult to regain trust.

There are some stories that should not be told. There are stories that probably should remain confidential. Leaders who have the well being of the organization foremost in mind, that is to say leaders who are *part of something bigger than themselves*, can *trust* their intuition with respect to which stories should be countered, nipped in the bud, or not told at all.

LEADERSHIP REFLECTION

Observations of natural systems show that there is strength in diversity because it results in novel approaches to problem solving. Diverse views can also result in dissonance—a normal component of growth. Wise leaders do not feel threatened by diverse views of reality and even seek out diverse views.

1. *Are you as a leader open to the stories that are circulating through your organization even when they diverge from your personal view of reality? How do you as a leader ensure that members of your organization feel comfortable sharing their ideas in an atmosphere that respects diverse points of view?*

2. *What are the "grande narratives" (Boje, 2001: 10) in your organization? Who tells the organization's grand narratives and how are they formed? What are stories within your organization that diverge from the "grande narratives"? What can the organization learn from those stories?*

3. *What underlying themes connect the stories you have identified above and what are the dominant narratives? Do members of the organization recognize that "there are at least two sides to every story," or is the dominant narrative "my way or the highway"?*

(See Appendix D (Storytelling guidelines) to hone storytelling skills)

SUMMARY OF PART III

When the Navy Chiefs were viewed as a "storied space," themes began to emerge from the stories they told that reflected leadership narratives, which in turn prescribed ways of behaving for the Chiefs as well as others. Different Chiefs, in different locations, at different times, repeated similar leadership narratives and practices. The leadership narratives prescribed ways of thinking and behaving that were sometimes contradictory, but when dissonance occurred there was evidence that the participants attempted to modify or change their narratives to better adapt to their environments.

The Chiefs' leadership narratives mimicked the narratives of the larger storied space of the Navy and the military. There was evidence that their narratives functioned as attractors—patterns that enhance or inhibit growth—and that they engaged in double loop learning—learning from interaction with their environment and others, and passing on lessons learned to others. There was ample evidence that Navy Chiefs use stories—as well as the more expansive narratives and discourse that surround them—to negotiate individual and collective views of reality that in turn influence behavior and adaptation to internal and external environments.

Chiefs, as a group, are the "old Yodas"—the keepers of the culture of the Navy. They are skilled at functioning in networks that reach within and across organizations as well as up and down the chain of command.

The following section further distills insights and discusses some of the implications of those insights. It also includes recommendations for tapping into the source of energy that story and narratives in organizations can be.

PART IV

ORGANIZATIONAL ADAPTABILITY AND CREATIVITY THROUGH STORY

Man is a teller of stories, he lives surrounded by his own stories and those of other people, he sees everything that happens to him in terms of these stories and he tries to live his life as if he were recounting it.

Jean-Paul Sartre, *The Words*

OVERVIEW

Sometimes a synthesis of several ideas comes closer to the truth than any one idea does individually (Goerner, 1999). When Navy Chiefs were viewed as a "storied space" (Baskin, 2008: 2) with a shared view of reality, leadership principles and practices emerged. The Chiefs use of story and narrative illustrated how they share knowledge and use that knowledge to negotiate ways of responding to their world. Applying paradigms developed from observations of natural systems to an exploration of the stories and narratives told by Navy Chiefs exposed insights into how narrative and story works in human organizations.

The findings discussed in this book corroborated the findings of other research studies that have found that narrative has implications for organizations as well as the individuals who comprise them. While a discussion of the leadership principles and practices of Navy Chiefs has particular relevance for people working with, and within, the Navy there are implications that go well beyond the Navy.

Chapter 11

THE IMPLICATIONS OF MINDFULLY ENGAGING STORY

There was evidence that Navy Chiefs use stories and narratives to make sense, give sense, cope and prescribe behaviors that were then used to guide their actions. Many stories accomplished all three functions—sense making, coping and navigating into the future—simultaneously. The following discussion summarizes how the stories of Navy Chiefs help them accomplish those functions.

STORIES AND SENSE MAKING

Story is the vehicle we use to make sense of our lives in a world that often defies logic.

Jim Trelease

Carl Weick (1995) coined the phrase "sense making and sense giving" to express the ability of humans to use narrative to interpret something that has happened—that is to say "make sense" of it—and pass on that interpretation—"give sense"—to others. Navy Chiefs use stories to make sense and give sense. They tell stories couched in more expansive narratives to resolve the dissonance inherent in their world. They make sense by creating—through story—shared interpretations of reality that result in leadership narratives and behavioral maps—called leadership practices herein.

Master Chief Cal made sense of how he as a young man who made many mistakes could succeed in the Navy in *Master Chief Cal's Story*. He used his personal story to give sense to others, communicating to them the leadership practice *live and learn*, implying that the people who work for him can make mistakes and succeed if they learn from their mistakes. When he commented that he did not think he "would make it in the Navy today," he was attempting to make sense of the dissonance presented by an environment that had changed since he was a young sailor. Through the emerging narrative that surrounded his stories he was reshaping his view of reality, testing a dominant narrative and prescribing ways of behaving in response to a new, emerging reality.

People use story to make sense of complex and sometimes confusing situations. In *Bow Planes Incident*, Chief Dante was making sense of how a person who was supposed to be a trusted and highly intelligent leader—the Captain of a submarine—could make a grievous mistake. Chief Dante used the story to communicate his view of effective leadership to his audience. In *Bow Planes Incident* Chief Dante was testing the leadership practices *live and learn, suck it up* and *trust*.

In *Dude What Happened*, Master Chief Hassan was making sense of a confusing situation involving a perceived injustice to a sailor; he was using the story to communicate the importance of reciprocal trust and he was calling upon his fellow Chiefs to be forthright even when they are experiencing the dissonance that is inherent in going against the grain. In *Dude What Happened?*, Master Chief Hassan juxtaposed, and wove a thread, between the practices *trust, suck it up,* and *take care of your people.*

Individuals create and sustain images of a wider reality in part to rationalize what they are doing. Navy Chiefs sometimes told "touchstone stories" (Boyce, 1995: 18) that enabled them to coalesce around a common and collective view of reality. *9/11 on the Connie* was an example of a touchstone story that coalesced individuals around the narrative *You are part of something bigger than yourself.* Master Chief Tracy used *9/11 on the Connie* to rationalize the need for the ship to return to sea after the terrorist attacks upon the United States in 2001. The "9/11" stories that followed in the group discussion reinforced Master Chief Tracy's point and confirmed the group's collective view of reality.

Navy Chiefs use story to organize and sequence their thoughts by compressing the timing of significant events, or stretching out the timing of events, to communicate understanding and create a desired effect. They use stories to create causal connections and sequences of action that did not exist when the events originally occurred, thereby retrospectively making sense and giving sense to their audiences. Senior Chief Sam used the story *AB or Wanna Be*, to explain how he came to be a successful Aviation Boatswain Mate. Senior Chief Tim crafted a story with causal connections and an exciting plot in *USS Kitty Hawk Mishap*, to make sense of the confusing events that took place when a jet crashed on the flight deck of an aircraft carrier. In *New Chief Brings Hope*, Senior Chief Dee compressed events that happened over an extended period into a short story that gave sense to others about how to succeed and be a good leader. All of these stories contained a moral or a point that connected them to other stories, and thereby made sense in the context of a larger narrative that emerged.

USING STORIES TO COPE IN THE PRESENT

We are lonesome animals. We spend all of our life trying to be less lonesome. One of our ancient methods is to tell a story begging the listener to stay and to feel.

John Steinbeck

Navy Chiefs tell stories that help them cope in the present. They tell stories that are cathartic and enjoyable such as *Dippy Sippy Donut Guy* and *French Fries*. They tell stories to create bonds within the group, such as *9/11 on the Connie,* and *Single Parent Deploys.* They tell stories that increase tension and decrease tension such as *USS Kitty Hawk Mishap, Bow Planes Incident, Good Morning Sticker* and *We're Gonna Ace It.* They tell stories to get attention, bolster their egos and feel good, such as *Big Papa in Waikiki* and *Bow Planes Incident.* They tell stories to help others cope such as *Stroking on his Deathbed,* and *AB or Wanna Be.* They use stories and the narratives surrounding them to actively resolve dissonance in the present. And they use story to prescribe ways of behaving—called leadership practices here—such as *suck it up, take care of your people, live and learn, head on a swivel, stay flexible, trust* and *do not take yourself too seriously but do take your work seriously.*

In the present, Navy Chiefs use story to give sense to those who are listening so that they can cope. For example, in *AB or Wanna Be* and *Stroking on his deathbed,* Senior Chief Sam and Senior Chief Tim were helping the young Aviation Boatswain Mates who worked for them feel good about their technical specialty even though the work was at times physically grueling and at other times tedious, boring or unglamorous. They were helping their subordinates feel proud of what they do and thus cope.

When events that happened in the past did not make sense, or when leadership narratives failed to enable sense making or sense giving, Navy Chiefs engaged in discussion surrounding the stories that reflected emerging narratives—or new ways of making sense of what happened. In this way the Chiefs were constantly reassessing their leadership narratives and behavioral maps to respond to new information and cope in the present. Master Chief Cal engaged an emerging narrative when he questioned if he would "make it" in the Navy today.

There are behavioral maps embedded within stories of Navy Chiefs that support the internalization of organizational norms and provide a means through which organizational behaviors are managed. Likewise, leadership practices, such as *suck it up, live and learn, stay flexible, take care of your people,* and *trust,* are embedded within the stories of Navy Chiefs and provide a means through which organizational behaviors can be managed. Other researchers (Hansen, 1993) have found that being assertive, being a team player, taking risks and personal commitment to the organization were highly valued qualities in the organizations they studied and linked to effective leadership. Navy Chiefs tell stories that reflect similar highly valued qualities. *Bow Planes Incident* and *Dude What Happened* reflect the value of assertiveness. *New Chief Brings Hope, Remember the Stark*, and *We're Gonna Ace It* reflect the value of teamwork. *9/11 on the Connie and USS Kitty Hawk Mishap* reflect the values of courage and commitment.

Other researchers (Hansen, 1993) have found that organizations have archetypal stories and that members of the same occupational culture tend to tell stories that are more alike than different. Navy Chiefs located in different locations at different times told stories that were more alike than different. The stories the Chiefs told reflected similar leadership narratives and practices. And the Chiefs told archetypal stories that functioned as cultural codes with implicit morals reflecting the shared values and belief systems of the organizational culture. The stories of Navy Chiefs contained morals that reflected shared values but they also expressed dualities as the Chiefs negotiated a balance between practices such as *taking care of your people* while expecting them to *suck it up* and *live and learn*.

USING STORIES TO NAVIGATE INTO THE FUTURE

Those who do not have power over the story that dominates their lives, the power to retell it, rethink it, deconstruct it, joke about it, and change it as times change, truly are powerless, because they cannot think new thoughts.

Salman Rushdie

Some of the stories the Chiefs told, such as *USS Kitty Hawk Mishap* or *Remember the Stark*, functioned like event simulators through which listeners could vicariously participate in the experiences of others from a safe distance. Leadership narratives prescribed behavioral practices for how to respond to similar events in the future.

Most stories, such as *Raised by Wolves, We're Gonna Ace It*, and *Page 10's*, performed multiple functions; they functioned to make sense of the past, thereby enabling the tellers and the listeners to cope with the present and proceed into the future. Some stories such as *Dude What Happened* served to inoculate against future misfortune by describing how to avoid situations that could be harmful to the individual, or the group.

THE DARK SIDE OF STORY

The most erroneous stories are those we think we know best—and therefore never scrutinize or question.

Stephen Jay Gould

Up to this point, stories and narratives have been portrayed as mostly positive, and potentially powerful, forces that, albeit difficult to tame, have been largely under-utilized. But it would be erroneous to conclude that story and narrative are either benign forces in organizations or the panacea for all organization ills. Stories can function as a release valve from the pressure of discomforting situations that cannot be handled directly, and they can identify threats to the organization that cannot be identified in other ways, but they also can be used to attack, or protect, individuals or groups of individuals inside or outside of the organization (Feldman, 1990). And any discussion of story in organizations would be incomplete without addressing the potential for its abuse—the dark side of story.

Some of the dark aspects of story and narrative have already been alluded to or addressed in previous chapters, but they merit repeating and further discussion here. As was discussed earlier, stories and narratives can be used by those in positions of power to marginalize individuals or groups of people. Organizational myths can be used by the dominant coalition in an organization to camouflage its power, make decisions in secret, and hide the results of those decisions (Boje, Fedor, & Rowland, 1982).

Stories can be used to identify scapegoats, internal or external, to the organization that, justifiably or not, are then saddled with blame for the problems the organization faces or the pain it has experienced. Members of the organization come to believe that once the scapegoat is eliminated their problems will be solved—and sometimes eliminating the scapegoat actually does enable the organization to mediate conflicts and contradictions that arise from changing circumstances, but often the problems persist after the scapegoat has been removed. Deflecting blame through scapegoat stories may be temporarily effective, but it does little to move the organization in the direction of novel solutions that will be more enduring.

Likewise, stories and narratives that are fabricated for the purpose of social or political control—propaganda—will eventually be exposed as false in a culture that values and protects rights of free speech. People will tell stories that reflect the world as they see it, and if that view of the world strikes them as more authentic than the fabrications—propaganda—of those seeking control, falsehoods will be exposed.

For centuries, select groups of people—those with the most power and influence—have decided how to interpret history. Often, in organizations, there are a small number of people—those with the most power and influence—who determine the content of the "grande narratives" (Boje, 2001: 10); they have had "narrative hegemony" (Boje, 2001: 7) over the stories that will be accepted as truth. But other stories have always existed. They are the stories told on the shop floor, or the "mess decks" in the Navy. In the post-modern world, there has been growing awareness of the other narratives within organizations, and an interest in reinterpreting history from different perspectives, resulting in increased creativity and adaptability of organizations.

INCREASED AWARENESS AND INTERPRETATION OF REALITY

We construct a narrative for ourselves, and that's the thread that we follow from one day to the next. People who disintegrate as personalities are the ones who lose that thread.

Hannah Arendt

The primary benefit of studying narrative in organizations is increased awareness and insight that can inform behavior and thus enhance adaptability. Examining story increases awareness of what an individual or group values as well as how they have interpreted reality, and how they continue to adapt their interpretation of reality to new information that is received. Interpretation of reality influences what the individual or the group sees as the truth.

Insights regarding the dynamic nature of story, informed by paradigms adopted from natural systems, help individuals and groups understand that perhaps their interpretation of reality is not the only, or the best, interpretation of reality. If members of a group are able to accept that their interpretation of reality may not represent the whole truth, they may be open to, and comfortable with, considering other views, and may even seek out other views to enhance the adaptability and creativity of their group. Healthy human social systems need both "guardians" and "synergists" (Goerner, 1999: 212). The guardians are naturally wary and protect the status quo—the system as it is, while the synergists experiment with change. Disaster results if either side is allowed to run amok. Guardians and synergists can use story and narrative to work through the natural dissonance presented by conflicting views and modify or change narratives and behavior that will enhance organizational effectiveness.

Chapter 12

PRACTICAL APPLICATIONS AND RECOMMENDATIONS

It is not the voice that commands the story; it is the ear.

Italo Calvino

The implications of the findings presented here for those seeking to cultivate effective leaders are that classroom leadership training is most effective when coupled with practical examples—that is to say, stories—derived from the students' own experiences or the experiences of others. Practical examples are more than case studies with a school-house

---

moral. Case studies can be effective stories but stories will be more effective and dynamic when they are derived from an individual's own experience. Cultivating leaders' ability to reflect upon their narratives as individuals, and the narratives of the storied spaces they occupy is the best way to internalize lessons that will lead to insights that will sustain the individual in deeper and more lasting ways.

Master Chief Quin commented that as a young sailor she learned enough about leadership in three months of duty in the Chiefs' Mess to sustain her entire career. Master Chief Tom commented, "To this day when I am faced with a problem, I often ask myself, 'What would Master Chief Bell do in this situation?'" Both of these highly successful Master Chief Petty Officers were drawing upon lessons they had internalized from people they admired, and worked with, over two decades earlier. They had ready reference to behavioral maps that were part of a broader narrative, some conscious and probably some pre-conscious, or even unconscious, that continue to influence their behavior to this day. They demonstrated that they have the ability to tap into a part of their storied space that allows for play, creativity and innovation, and to imagine a dialogue of sorts with someone they have not spoken with or seen in years. The capacity of humans to use their imaginations in this way is powerful and has huge implications for leadership training as well as organizational creativity and adaptability.

STORIES AS SIMULATORS— OPERATIONALIZING INTUITION

We naturally strive to operationalize knowledge.

Robert Chia

Beyond leadership training, stories have the capacity to function as simulators, allowing individuals to experience dangerous situations from a safe distance and practice behaviors for how to effectively handle such situations when, and if, they are encountered in the future. I cannot say for sure how stories such as *USS Kitty Hawk Incident*, or *Remember the Stark*, will effect the behavioral maps of the sailors who heard those stories, but other researchers have found evidence that such stories do affect behavior and can even serve to "operationalize" intuition (Chia, 1998). Confirming such findings could be a tremendous boon to the Navy and the military in general, as the military searches for safe and less costly ways to effectively train people for dangerous situations.

Weick (1993, 1995) analyzed, and wrote extensively about, the narratives of fire fighters and others who worked in high-risk occupations. The stories he analyzed were similar to *USS Kitty Hawk Mishap*. Weick suggested that recollection of a story can slow down escalation in a frightening situation, thereby slowing the rate at which pressure builds. Weick suggested that a well-rehearsed story can help people simplify the task at hand and help them tolerate more pressure. Stories can reduce the element of surprise, and once the pressure is reduced and the pace is slowed people can be more attentive to both to the central and peripheral cues in the environment. Weick also suggested that while stories may help people manage pressure and improve sense making during emergencies they may be even more helpful in the prevention of emergencies, because dealing with imagined threats for obvious reasons is far less dangerous than dealing with actual threats. Weick said that stories can safely prepare people for crises by rehearsing both plausible and sometimes seemingly implausible circumstances (Weick, 1993, 1995).

Chia (1998) suggested that narrative plays a role in the development of "the Intuitive Method" (p. 358). Through communicating lived experiences, story can maintain a necessary level of tension that can sustain complex adaptive systems operating at the edge of chaos, or what Lewin and Regine (2001) described as the "zone of creative adaptability" (p. 28). Accident stories may serve to sustain a complex adaptive system by operationalizing intuition (Chia, 1998). Gabriel (2000, 2004) said that stories about accidents could serve to increase tension and inoculate listeners against future misfortune.

We naturally strive to "operationalize" knowledge (Chia, 1998). That is to say, we try "to translate concepts and ideas into measurable forms in order to render them more amenable to cognitive manipulation" (Chia, 1998: 345). He suggested that story may be an effective way to communicate a deeper more complete meaning by allowing humans to vicariously experience someone else's world. In this sense, using story may be one way to enhance intuition. Members of the military, like Navy Chiefs, intuitively appreciate the value of "mishap" stories and have been using them for years both formally and informally in training to prepare others to effectively manage emergencies.

If story can serve to enhance intuition and "inoculate" (Chia, 1998) against misfortune it could be of tremendous value in the Navy and to people working in other high-risk organizations and it needs to be explored further.

BRIDGING THE GAP

Before each class at the Senior Enlisted Academy the Chief Petty Officers recite the Sailor's Creed. Every Navy recruit is given a copy of the Sailor's Creed and is required to commit it to memory. The Sailor's Creed supposedly embodies the essence of what it means to be a sailor. It is as follows:

> *I am a United States Sailor. I will support and defend the Constitution of the United States of America and I will obey the orders of those appointed over me. I represent the fighting spirit of the Navy and all who have gone before me to defend freedom and democracy around the world. I proudly serve my country's Navy combat team with Honor, Courage and Commitment. I am committed to excellence and the fair treatment of all* (Chief of Naval Operations Blue Ribbon Panel, 1993).

The words "honor, courage and commitment" are emblazoned on everything from posters to mission statements in the Navy. I expected to hear stories about honor, courage and commitment and I did hear such stories, but rarely were the actual words "honor", "courage" and "commitment" used in the stories and narratives told, and those three words were never used together in any of the stories collected for this book. When asked what was meant by the words "Honor, Courage and Commitment," the Navy Chiefs—who otherwise were never at a loss for words—were mute. Yet they told many stories that contained examples of honor, courage and commitment. Senior Chief Tim displayed courage in *USS Kitty Hawk Mishap*. Master Chief Tom displayed commitment in *Page 10's*. Senior Chief Sam displayed honor in *AB or Wanna Be*. But the actual words "honor, courage and commitment" were not used in those stories. Honor, courage and commitment are abstract concepts, it takes a story to paint a picture of what they really mean.

Likewise, an overarching need for "humility" is emphasized in The Chief Petty Officers' Creed that is seared into the brains of all Chiefs the day that they are initiated into the brotherhood of Chief Petty Officers. As the following excerpt from The Chief Petty Officers' Creed illustrates, the Chief Petty Officers' initiation is designed to be a lesson in humility.

You were subjected [during the initiation] to humiliation to prove to you that humility is good and great, a necessary attribute which cannot mar you, and in fact strengthens you ... you will be caused to suffer indignities, to experience humiliation far beyond those imposed upon you today. Bear them with the dignity and with the same good grace which you bore them today! (Navy Advancement Study Guide, 1998).

But humility is an abstract concept. The excerpt from the Chief Petty Officers' Creed cited above reflects the leadership narratives *You are part of something bigger than yourself* and the leadership practices *live and learn, suck it up, trust* and *do not take yourself too seriously but do take your work seriously*. None of the participants used the words "humility" or "humiliation" in the stories they shared, or in the discussions that surrounded the stories; and while the participants told many stories that involved self-deprecating humor, none of the stories told were about "humiliating" experiences. The stories about potentially humiliating experiences were told as epics, reflecting the protagonist's ability to endure, or perhaps *suck it up*. Actual stories were needed to paint a picture of what humility looks like.

Those seeking to guide organizations often attempt to simplify information by reducing it to abstract words or phrases that fail to communicate the meaning, or paint a picture, that bridges the gap between the abstraction and practical application. Organization vision and mission statements are an example of abstract phrases that often fail to paint a picture of what the values outlined look like in action. This is not to say that organizational directives such as mission statements are ineffective; they probably mean a great deal to the people who developed them, but unfortunately they do not mean as much to those who were not involved in the interplay between stories and the narratives surrounding them that it took to create them. As a result, once they are promulgated, circulated, or posted, they become lifeless artifacts of the workshop that took place to produce them. Stories can contain rich complexity and prescribe behavioral maps suited to a wide variety of situations. Leaders can consciously work to identify stories that humanize abstract thoughts and bridge the gap between abstractions and how to practically apply those abstractions.

The words "honor, courage, commitment" in the Sailor's Creed and the emphasis on humility in the Chief Petty Officer's Creed are examples of abstract concepts—they are not concrete examples of behavior; therefore, by themselves they do not prescribe easy to follow behavioral maps. Similarly, the other leadership principles distilled and described in this book do not transmit meaning very well by themselves but when coupled with stories they do. The concepts of "honor, courage, commitment" and the principles used to characterize leadership narratives are much richer when they are combined with examples—stories—that illustrate their meaning.

Story can bridge the gap between quantitative knowledge and human experience, resulting in a more useful application of knowledge to real world challenges. Like the Challenger and Columbia space shuttle accidents discussed in Chapter 1, quantitative data requires narrative interpretation to be useful. And human interpretation is based on individual and group interpretations of reality that have been formed over time through the interplay of quantitative data and narrative interpretation.

IMAGINING A POSITIVE FUTURE

The history of a soldier's wound beguiles the pain of it.

Laurence Sterne

If humans can find guidance in their recollections of stories about former mentors, and rehearse for plausible and implausible scenarios through stories, they should also be able to imagine narratives about positive future outcomes that can guide behavior to achieve those outcomes. Emerging narratives are like dreams. They are the preconscious attempt to resolve dissonance, and once articulated they are an attempt to interpret and make sense. It has been suggested that through story humans can "imagine new futures, act on those stories and change the world so that they can realize such futures" (Baskin, 2008: 2). The implication is that through narrative and emerging narratives humans can unleash the creativity of their imaginations by turning their dreams for a positive future into stories and developing behavioral maps for how to achieve them.

People and nations, around the world, are constantly dealing with challenges and problems such as war, natural disasters, economic uncertainty, personal hardships and strife of one sort or another. As I was writing this book, there were more soldiers, sailors, airmen and marines returning from war than at any time since the Vietnam War. People around the world are constantly struggling to make sense, cope, and successfully navigate into the future. People and organizations suffering hardships can benefit from an understanding of the dominant narratives that serve to guide their behavior and the behavior of the storied spaces they occupy. Stories that function as limiting attractors—limiting or blocking adaptability—can be identified and changed, or modified, to enhance the adaptability of individuals and organizations. Self-reflection and self-awareness can help people identify personal stories that will enable them to cope and thrive.

For example, members of the military transitioning to civilian employment, can identify personal stories that illustrate the traits that make them excellent candidates for jobs, such as the personal epics about how they have grappled with challenges and prevailed, the personal comedies about the mistakes they have made and what they have learned from those mistakes, the stories about their commitment to *something bigger than themselves*, the stories about their stamina and ability to *suck it up* when the going gets tough, the stories about their commitment to the team and how they have successfully collaborated with others. Such personal stories illustrate values and traits that will make them desirable additions to any organization.

Likewise, those assisting military members in transition can benefit from understanding the narratives that have informed the values and behavior of military members, and the organizations they have been a part of, as they guide them through transitions. Understanding personal and organizational narratives can help those assisting transitioning military members identify traits they have honed in the military—such as self-sacrifice, teamwork, diligence and stamina—that make them appealing candidates for jobs outside of the military.

Some military members harbor personal tragedies that they are unwilling, or unable, to share with others within the storied spaces of their organizations for good reason. Some bear deep emotional wounds and scars. They will not be able to fully function as healthy members of organizations, or society at large, until they grapple with those emotional wounds and make sense of what caused them. Mental health professionals, spiritual counselors, family members and close friends are in the best position to co-story with such individuals, helping them make sense so that they can cope and successfully proceed into the future. The organization—the storied space—can assist by *taking care of people* and ensuring that individuals who are suffering from emotional wounds receive the mental health assistance they need and that is available to all military members.

It is estimated that as many as one in four homeless on the streets of America are veterans (Gamache & Tessler, 2001), and women veterans are at a higher risk for homelessness than their non veteran counterparts (Gamache, Rosenheck & Tessler, 2003). Most homeless veterans are suffering from some sort of mental or emotional illness and are self-medicating in an attempt to escape their pain. No doubt, some veterans had undiagnosed mental or emotional prior to entering the military, but whether they had pre-existing conditions or not, as veterans they are legally entitled to treatment through the Veterans Administration. Why do they not avail themselves of such services? Does a dominant narrative persist among veterans that those who avail themselves of mental health services through the Veterans Administration are losers? Is there a dominant narrative in American culture that those who require mental health counseling are merely weak and not really ill? Is there a dominant narrative in American culture that mental health counseling is a futile endeavor?

Military leaders as well as leaders in organizations outside of the military can counter such narratives by not stigmatizing individuals who might benefit from the intervention of mental health professionals, underscoring the organizational value that *we are part of something bigger than ourselves*. When one member of the organization or a sub group of the organization is suffering the rest of the organization will feel the pain. It is in the best interest of the organization to *take care of its people* so that they can be fully productive members once again.

Schein (2006), who patterned his work with organizations on the psychotherapy he did with repatriated Korean War veterans suggested that drawing out the strengths of an organization through eliciting its stories is not unlike drawing out the strengths of individuals through eliciting their personal stories. He suggested that eliciting organizational narratives was an effective way for consultants to help organizations apply their strengths to their problems and successfully grapple with future challenges. He said,

The more I examined process consultation and observed my own behavior as a consultant, the more I realized that what consultants do is very akin to therapy, but this formulation is not acceptable to most managerial clients.

Organizational pathologies of all sorts are evident whenever one gets into client situations but in working with organizational cultures one must learn to use the metaphors and linguistic categories that make sense to them and enable them to save face and avoid defensiveness. I realized that the best kind of therapy draws on personal and culture strengths even though the process is triggered by pathology, weakness or problems. The consultant/therapist must learn to draw out the strengths in the culture and show how they can be used to solve the problems facing the organization (Schein, 2006: 297).

Organizations can use their stories to tap into the strengths of their culture and begin to grapple with the "organizational pathologies" that afflict all organizations. Leaders can function as "organization therapists" by using individual and organizational stories to make sense of where an organization has come from and what it is today and paint a picture through story of what an organization can be in the future.

Finally, the storied space needs to trust, that the values espoused in dominant narratives are conducive to positive future outcomes. Baskin (2008) suggested that the critical difference between humans and other animals, "is the ability of human beings to tell stories, and change the world so that they can realize such futures" (p. 2).

RECOMMENDATIONS

Preach the Gospel at all times, if necessary use words.

Saint Francis of Assisi

There is a time and a place for everything, including silence and action. Saint Francis of Assisi is said to have advised his followers to, "Preach the Gospel at all times, if necessary use words." The stories are always present but they do not always need to be articulated. Indeed, the articulation of story has its time and place. That being said, I hope that those who read this book will not write off its message as fluff that merely reinforces the time honored toastmasters' principle of starting all speeches with a joke or anecdote. I hope I have opened the reader's eyes to the ubiquitous presence of story—whether articulated or not—and its power to motivate, explain and energize humans in unique and dynamic ways.

I am recommending that leaders get in touch with stories that have made them who they are, and the stories that have made their organizations what they are. I am suggesting that leaders explore the meaning of those stories. I am recommending that all members of organizations think about how they might be use their stories to communicate meaning. I am suggesting that when leaders are faced with communicating the meaning of complex or abstract ideas that they think about how they might use a story. I am suggesting that leaders dialogue with the members of their organizations to identify the stories and narratives that have made the organization what it is. Once the organizational stories have been identified, the organization can make sense of why it is where it is today and begin to make necessary adaptations to successfully proceed into the future—imagining new stories that will enable the organization to prosper. This is not an easy exercise. It is difficult to dig deep inside ourselves and make sense of why we are who we are, but once we have done so we can begin to understand so much more.

I am recommending that leaders listen to the stories that they hear every day, in conversations, in the news, in books, in meetings. I am suggesting that they think about how those stories reflect a particular view of reality. I am recommending that leaders ask the people in their organizations to tell their stories and then listen carefully. People love to tell their life stories and the careful listener will begin to build a repertoire of stories that they can then use when they need to illustrate a point, teach a lesson, inspire the team, or sooth a soul.

CONCLUSIONS

Earlier in this book I explained that I had a hunch—a theory—about Navy Chiefs. I had a hunch that story and narrative had something to do with why Navy Chiefs are effective leaders. I also had a hunch that as a group Navy Chiefs functioned unlike other groups in the Navy and I set out to explore that hunch. I observed that Navy Chiefs were more cohesive and more cooperative than competitive with each other. I observed that they networked with each other across organizational lines as well as up and down the chain of command to get work done in ways that differed from other groups in the Navy, such as the "Wardroom" (the Commissioned Officers) on the ship. I observed that they functioned more like a natural, self organized system, than a machine. The stories and narratives of Navy Chiefs confirmed my hunches. Leadership principles and practices were uncovered in the stories and narratives of Navy Chiefs that illuminated the rich source of energy that story can be as it travels through an organization connecting individuals in a web like fashion with Navy Chiefs functioning as nodes in that web.

The conclusion is that story and narrative are as natural and integral to how work gets done in organizations as breathing is to human life—and they are just as taken for granted. Story and narrative in organizations are powerful drivers of behavior that are under-utilized and poorly understood. Following the example of Navy Chiefs, leaders in other organizations can mindfully engage story, and the more expansive universe of narrative that surrounds it to enhance the adaptability and creativity of their organizations in a fast paced, ever-changing world.

Ultimately, the discovery of truth is the goal of research. Like a ship sailing through a heavy mist, the truth appears at times with striking clarity but in the next moment is shrouded once again in mist. It was my intention to uncover some modicum of truth by allowing the stories and narratives of Navy Chiefs to speak for themselves to the extent that they can in a written replication and interpretation. The readers will no doubt have their own interpretations that will differ from mine, but I remain convinced that the truth lies within the stories waiting to be revealed to those who seek it. Therefore, the following *Story of Truth* is offered as a concluding comment.

THE STORY OF TRUTH

(a traditional Jewish folk tale, author unknown)

Truth walked into a village. The local inhabitants started cursing at him. Spewing epithets, they chased him out of the village. Truth walked along the road to the next town. They too spit at him and cursed and spewed epithets driving him out of town. He walked, lonely and sad, down the empty road, until he reached the next town, still hoping to find someone who was happy to see him who would embrace Truth with open arms. So he walked into the third town, this time in the middle of the night, hoping that dawn would find the townsfolk happy to see Truth with dawn's light. But as soon as the townsfolks' eyes lit upon him they ran to their homes and then came back throwing garbage at him. Truth ran off, out of town, into the woods, and after crying and cleaning off the garbage returned to the edge of the woods when he heard laughter and gaiety, singing and applause. He saw the townsfolk applauding as Story entered the town. They brought out fresh meats and soups and pies and pastries and offered them all to Story who smiled and lavished in their love and appreciation. Come twilight, Truth was sulking and sobbing at the edge of the woods. The townsfolk disdainfully ignored him, but Story came out to see what the story was. Truth told Story how all the townsfolk mistreated him, how sad and lonely he was and how much he wanted to be accepted and appreciated. Story replied, "Of course they all rejected you," Story looked at Truth, eyes a bit lowered to the side, and said, "No one wants to look at the naked Truth." So Story gave Truth brilliant, beautiful clothing to wear, and they walked into the town together, Truth with Story, and the townsfolk greeted them with warmth and love and appreciation, for Truth wrapped in Story's clothing is a beautiful thing and easy to behold. And ever since then Truth travels with Story and they are always accepted and loved. And that is the way it was, and the way it is, and that is the way it always will be (Henshall, 2005).

APPENDIX A

A VERY BRIEF EXPLANATION OF COMPLEXITY SCIENCE THEORY—SOMETIMES CALLED COMPLEXITY THINKING, OR THE SCIENCE OF EMERGENCE

Complexity Science theory explains that the world is made up of complex adaptive systems, or entities, that have the ability to self organize and maintain themselves—a process that is sometimes referred to as autopoeisis (Maturana & Varela, 1987). Viewed through a Complexity Science lens, organizations and the people who work in them function like complex adaptive systems—that is to say they have the ability to maintain themselves. A Complexity Science view has challenged the mechanistic model that has been used in the past to explain how work gets done in organizations. Simply put, the mechanistic model—sometimes described as "reductionist" or "linear"—popularized by Frederick Taylor in the early part of the 20th century fails to adequately explain how human social systems, such as organizations truly work (Goerner, 1999).

Waldrop (1992) used the term "Science of Emergence" to capture the unfolding, coming into being, dynamic of this diverse, interdisciplinary theory. But in the event this subject inspires a desire on the part of the reader to explore the topic further, the terms "Complexity Science theory" and "Complexity thinking" have been used in this book.

Complexity theory does not have a clear historical beginning. The term has been used by mathematicians, world-wide, for a long time as a theory of classifying problems based on how difficult they are to solve. Contemporary applications of complexity theory have their roots both in the complexity of mathematics and in chaos theory, which in turn has grown out of the work of many mathematicians and scientists in a variety of disciplines. Complexity thinking is not to be confused with systems thinking which is more aligned with hierarchical Newtonian views. The Santa Fe Institute, which is dedicated to the study of complexity theory, has promoted the study of the subject not only as a mathematical theory or a science but also in a broad, multidisciplinary context (Bloch, 2004, 2005; Bloch, Henderson & Stackman, 2007; Bloch & Richmond, 1998; Stackman, Henderson, Bloch, & 2006).

A number of researchers have used a complexity theory framework to analyze human interactions and human social systems (Baskin, 1998; Bloch *et al.*, 2007; Boje, 1991; Chia, 1998; Stacey, 1996, 2000; Wheatley, 2006). Viewed through a complexity thinking paradigm, organizations, and the people who work in them, function like complex adaptive entities (Baskin, 2008; Bloch, 2005; Bloch *et al.*, 2007; Boje, 2001; Chia, 1998; Gleick, 1987; Kauffman, 1995; Lewin, 1992; Lewin & Regine, 2001; Stacey, 1996, 2000; Wheatley, 2006).

But there are two researchers—Baskin and Boje—who overtly focused on a combination of narrative theory and complexity theory in their study of organizations—an approach that was adopted in this study. Both Boje (2001) and Baskin (2008) underscored the point that narrative in organizations functions in a non-linear fashion, adapting, sometimes changing and morphing to adjust to ever changing environments and conditions. They both found people in organizations use narrative to create meaning and a shared view of reality (Baskin, 2008; Boje, 1991, 1995, 1998, 2001).

APPENDIX B

A BRIEF HISTORY OF NARRATIVE AND STORY

The ancient Greek literati were the first to consciously examine the composition of narratives through their identification of themes in myths and legends. Much of the early foundational work in narrative analysis followed the pattern established by the ancient Greeks by identifying common themes, threads and morals in myths and legends. Greek myths, legends and fables, communicated the values and ideals that influenced the cultures that followed and evolved into the underpinnings of Western Civilization. But it was not until the 20th century that researchers and theorists from a variety of disciplines started to examine how and why story and narrative work the way they do.

Joseph Campbell found common themes in the myths of a wide variety of cultures that he believed defined universal truths. He was particularly interested in the idea of the hero archetype and suggested in his seminal work *Hero with a Thousand Faces* that myths about heroes can serve as role models as readily as real life mentors serve as role models. In *The Masks of God* Campbell explored universal themes in world religions. Campbell's work served as inspiration for novelists and filmmakers in the later part of the 20th century including George Lucas, maker of the *Star Wars* films (Campbell, 1973, 1988).

The child psychologist Bruno Bettelheim analyzed the emotional and symbolic importance of folk and fairy tales from a variety of cultures. He found common themes and morals in folk and fairy tales that allowed children to grapple with fear in remote and symbolic ways, thereby enabling them to better cope with fearful situations when they encountered them in real life.

Variations on The Brothers Grimm *Little Red Riding Hood* can be found in other cultures, the Chinese version casts a tiger in the role of the Big Bad Wolf and the Muslim version casts a boy rather than a girl in the role of the protagonist because a girl would never be sent off to Grandmother's house by herself (Bettelheim, 1970).

The psychoanalyst Carl Jung suggested that common themes in popular myths reflect the constantly repeated experiences of humanity. Jung explored Eastern and Western philosophy, dreams, art and mythology identifying psychological archetypes that underscored his theories about what he called the "collective unconscious." Perhaps best known for his dream theory, Jung theorized that the language of dreams gives us a glimpse into the unconscious mind in the liminal space between conscious and unconscious where symbols are unique and personal. A symbol used to represent a theme in one dream may mean something entirely different in another dream or may mean something entirely different to another person. Jung's study of dreams suggests that in both conscious and unconscious states of mind humans are constantly striving to make sense and deal with dissonance in their perceptions of the world. He theorized that what humans cannot make sense of consciously, they deal with unconsciously through their dreams. Jung suggested that people can deconstruct their dreams thereby employing them consciously to resolve dissonance and solve problems. He contrasted dreams with conscious story, but he suggested that dreams can be turned into useful narratives that can help individuals cope and move in a positive direction (Jung, 1964).

20th century philosophers such as Paul Ricoeur (1984) legitimized the function of "story as text" as a way of interpreting deep philosophical meaning in humans and their institutions. And anthropologists like Margaret Mead used ethnography, that is to say the telling of the story of a human social group through their own stories, as a valid way of interpreting cultural artifacts. (Mead, 1934; Ricoeur, 1991, 1992)

In the later part of the 20th century and early part of the 21st century, academic researchers and business consultants who were interested in exploring organizational cultures and the nature of leadership in organizations turned to story as a rich repository of meaningful data that could not be collected in other ways. Peters and Waterman collected stories of successful organizations and leaders culling out universal themes that evolved into the *McKinsey 7-S framework* of organizational excellence. Their seminal work—

In Search of Excellence—was the first book in the popular business press to relay in a story format the lessons they gleaned from the stories of successful companies (Peters & Waterman, 1982). Jim Collins distilled principles from the stories he collected from his work with successful organizations in *Good to Great* (2000).

The industrial psychologist Edgar Shein, who explored the notion of organizational culture and in his case studies of successful leaders, found that anecdotes about the founders and leaders in organizations took on a mythic quality and served as guideposts for future actions. Like Peters and Waterman, Schein approached the study of organization culture through narrative. Through his interviews with leaders and the people who worked with them he was able to identify themes, and the characteristics of successful leaders. While Schein did not focus specifically on the role of story in his studies, he included numerous examples of anecdotes in his findings, and concluded that stories told by leaders, and about leaders, played a crucial role in transmitting organization values and guiding the direction of the organizations and leaders he studied (Schein, 1983, 1985, 1990, 2006; Schein & Barker, 1961; Schein & Bennis, 1965).

Until the later part of the 20th century it was largely assumed that organizations reflected their dominant national or ethnic cultures, but industrial psychologists like Schein and Organizational Development practitioners documented the differences in the success rates of seemingly similar organizations that pointed to the existence of distinctly different organizational cultures and subcultures. Story is only one way that organization culture manifests itself, but story has proven to be a particularly rich artifact for those seeking to understand the underlying values and assumptions within organization cultures and subcultures.

In *When Jesus Came to Harvard*, Harvey Cox (2004) described a business school course on morals and ethics that he taught at Harvard based on the parables of Jesus Christ. Cox created this course in the wake of the Enron scandal and other accounts of moral and ethical debauchery in American business. Cox suggested that everyone raised in Western culture— whether they are Christian or not—is familiar with parables from Christian Scripture and that Jesus Christ provided an example of moral reasoning through the stories he told and enacted, "He (Jesus) forced people to think for themselves using stories that we might now call case studies…he used his stories to make people grapple with moral issues but not in a vacuum" (Cox, 2004: 27-28). Most of us understand what is meant when someone is described as a "Good Samaritan" or a "Prodigal Son." Cox suggested that as a model for moral reasoning the teachings of Jesus are just as valid as the teachings of Aristotle or Plato.

In summary, the organization research using story and narrative has been built on a larger foundation of work on narrative especially in the areas of myth and legend conducted by folklorists, anthropologists, sociologists, psychologists and linguist's who have proposed that story may be a way of uncovering insight into people and how they function in a complex, sometimes chaotic world.

APPENDIX C

SNAP SHOT OF PARTICIPANTS

Rank, Pseudonym	Gender	Ethnicity	Technical Specialty
1) Senior Chief Dee	Female	Caucasian	Logistics
2) Master Chief Mike	Male	Caucasian	Construction
3) Master Chief Hassan	Male	African American	Sonar
4) Senior Chief Brian	Male	Caucasian	Operations
5) Senior Chief Carl	Male	Caucasian	Communications
6) Master Chief Jeff	Male	Caucasian	Weapons
7) Master Chief Eric	Male	Caucasian	Electrical
8) Chief Dante	Male	Hispanic	Engineering
9) Senior Chief Shane	Male	Caucasian	Engineering
10) Senior Chief Dave	Male	Caucasian	Aviation
11) Senior Chief John	Male	Caucasian	Sonar
12) Senior Chief Tracy	Male	Caucasian	Aviation
13) Master Chief Tom	Male	Caucasian	Personnel
14) Senior Chief Bob	Male	Caucasian	Sonar

Table 1 *Senior Enlisted Academy Participants*

Rank, Pseudonym	Gender	Ethnicity	Technical Specialty
1) Master Chief Cal	Male	African American	Aviation Boatswain
2) Chief Saul	Male	Hispanic	Aviation Boatswain
3) Senior Chief Tim	Male	Caucasian	Aviation Boatswain
4) Chief Sara	Female	Caucasian	Hospital Corps
5) Senior Chief Evers	Male	African American	Aviation Boatswain
6) Senior Chief Sam	Male	African American	Aviation Boatswain
7) Chief Bryant	Male	Caucasian	Aviation Boatswain
8) Senior Chief Vela	Male	Hispanic	Aviation Boatswain
9) Senior Chief Jerry	Male	Caucasian	Communications
10) Senior Chief Jeffers	Male	African American	Aviation Boatswain
11) Chief Brent	Male	Caucasian	Aviation Boatswain
12) Chief Rolf	Male	Caucasian	Boatswain
13) Master Chief Arcelo	Male	Asian	Mechanical Repairs
14) Senior Chief Pay	Male	Hispanic	Boatswain
15) Master Chief Joe	Male	Caucasian	Engineering
16) Master Chief Ray	Male	Caucasian	Administration
17) Master Chief Quin	Female	Caucasian	Personnel
18) Chief Cord	Male	Hispanic	Aviation Control
19) Senior Chief Tray	Male	Caucasian	Aviation Boatswain
20) Senior Chief Tully	Male	Caucasian	Aviation Logistics

Table 2 *Ship Participants*

APPENDIX D

STORYTELLING GUIDELINES AND TIPS

There are numerous books, tips and guidelines containing suggestions for how to hone the craft of "performance storytelling." Performance storytelling is like acting, and honing storytelling ability is not unlike honing acting ability. Like acting there are classes and workshops given throughout the country that are dedicated to cultivating professional storytelling skills (go to: www.storynet.org for classes near you). But most of us are not, nor do we aspire to be, professional storytellers. Therefore, the following guidelines are not intended for the professional storyteller, rather they are intended for the reader who wants to use stories to make sense, cope, entertain, persuade, make a soul to soul connection with others and explore a world of ideas that are emerging, evolving, revelatory and powerful vehicles for complex meanings and truth.

Throughout this book I have urged the reader to recall the personal stories that have made them who they are. Or I have urged the reader to elicit personal stories from others. These are the types of stories that are revealed and shared daily as we go about the business of living and working with others. The following guidelines are intended to serve those who want to tap into the power of story in their day-to-day lives.

Personal stories usually arise in one of two ways: 1) An event occurs that stimulates a desire, or need, to articulate a story for reasons that may not be crystal clear initially to the teller or the audience; or, 2) someone wants to communicate meaning of some sort—perhaps a lesson or a bit of wisdom—and offers a story that captures the meaning they are attempting to communicate. Many examples of such personal stories are included in this book and after contemplating the reflections at the end of each chapter the reader hopefully started to recognize their own story repertoire and added to it.

What follows are some suggestions for how the reader can use their own stories, or stories collected from other sources, to communicate and engage with others. Take one of the stories from your repertoire and explore the following:

1. First think about what makes the story interesting. Why do you like the story? What parts of the story do you especially like?

2. Next think about what points, lessons, truths or messages are contained in the story. Does the story reveal the elephant in the room? Does the story disclose a secret or offer a gem of wisdom?

3. What do you think the story says about you, your values, your character, or the values of the organization? Are these meanings that you are trying to communicate?

4. Does the story make sense of something that did not make sense initially? Does the story provide a map for how to behave in the future? Does the story reduce tension, create tension, entertain or create a bond?

5. Does the story make something ordinary seem extraordinary? Conversely does the story make something extraordinary seem ordinary?

6. What is the most important thing about the story? What do you want the audience to take away?

7. What secondary themes, or "take aways," does the story contain?

8. What is the story's plot? What conflict or dissonance must be resolved? The plot is the glue that holds the story together. What problem, or problems, are the characters in the story struggling with? What puzzle, or puzzles, must the characters resolve?

9. What kind of audience might be interested in this story or how could it be adapted to different audiences?

Once you have answered the questions listed above you can begin to contemplate the practical aspects of how to tell and use your story. Remember a story needs a beginning that sets the stage for the action and places it within the context of the discourse that preceded it. A story also needs a middle that develops the plot and an ending that sums up the plot resolution. If it is a personal story take yourself back to the time and place the story describes, relive it, feel it, see it. If it is someone else's story try to imagine what it must have felt like to be there. If it is a well known tale, myth or fable think about how it enhances the meaning you want to communicate. Let your voice, actions and presence be a vehicle for the authenticity of the story. Establish eye contact with your audience. Speak to them as individuals, to create a sense of intimacy.

1. Introduce the story to get your audience's attention. Let them know that you are about to tell a story. The introduction brackets the story in the discourse that surrounds it. If the story is told in the context of a group conversation, the group's discourse ceases while you tell the story—you are essentially asking the audience's permission to "take the floor." If the audience is receptive and willing they will allow you to do so. Some typical introductions include: "So there we were…" or "That reminds me of a story…" or "Once upon a time…"

2. Describe the setting. Paint a picture for your audience. Develop the context. Help them see it and feel it. Arouse the audience's senses… "It was a dark and stormy night…"

3. Give your audience a feel for the characters, describe them, communicate their character traits…"he was a salty old master chief, who smelled like diesel fuel and smoked two packs of Camels a day," or "he had a coffee mug that hadn't been washed in years because washing it would 'ruin the flavor.'"

4. Describe the plot. What problem or puzzle needs to be resolved? What doesn't make sense? Through description and dialogue build tension as the character, or characters, struggle with the problem or puzzle. Speed up or slow down the telling of the story to create tension as it builds to a climax.

5. Pause strategically at the climax of the story right before the problem is resolved, or before telling the "punch line" to underscore the resolution of the problem or puzzle.

6. Conclude with a moral, if appropriate, that puts the story into the context of the discourse that surrounds it. For example: "Even though I worked with him for less than a year—and that was 30 years ago—to this day when I am faced with a problem I often ask myself 'What would Larry Bell do?'" Or "And ever since then, Truth travels with Story and they are accepted and loved, because Truth wrapped in Story's clothing is a beautiful thing and easy to behold," or "So oft in theologic wars…"

Finally, think about where and how you could use this story. Stories can be expanded or contracted to fit the moment or the audience. Some stories can be communicated with a single sentence in a few seconds while others require development and benefit from elaboration. Know your audience and make the story relevant for them. The same story may be used to communicate different meanings in different settings, with different audiences, at different times. You may put away a story for years but bring it out later at an appropriate moment. Read your audience as you tell the story just as you might read your audience if you are giving a speech, carrying on a conversation, or teaching a class. Trust your judgment and senses. Do not be afraid to stop and ask the audience if the story is striking a chord with them and adjust the discourse as necessary. Maybe it is not the time or the place for a particular story or any story at all— and that is a story in itself!

Through story you are inviting your audience to take a journey with you and you are offering them a reward for taking that journey with you—perhaps a lesson, an insight, or a bit of fun—and that is why they have given you the floor. You owe it to them to deliver. So if you sense you are not striking a chord do not take it personally, be self-aware enough to adjust your message or delivery as necessary. Remember the goal of storytelling is to communicate meaning—let the meaning be foremost.

Finally, make sure there is congruence between the points in the stories you tell and your behavior, or if there is not congruence point out why. As was discussed earlier in this book, nothing will unmask an egotist, phony or poser faster than a story. Most importantly, enjoy the sacred magic of storytelling and practice it often because it is our stories that give life purpose and meaning.

Story telling resources

- Stephen Denning on story in organizations;
- Daniel Pink on developing "right brain" capabilities including storytelling;
- Doug Lipman and Ruth Sawyer on improving the craft of storytelling;
- Terrence Gargiulo on evoking stories in organizations;
- Robert McKee on story development and screen writing;
- Harvey Cox on using story to teach morals and ethics;
- Joseph Campbell on the power of myth from ancient times to the present, and;

- Christina Baldwin on making sense of our lives through our personal stories.

The National Storytelling Network for regional conferences, concerts, workshops and numerous other resources that showcase the craft of storytelling and foster development of its use in a multitude of ways

APPENDIX E

AUTHOR'S BIAS, BACKGROUND AND OTHER LIMITATIONS

Students of organization studies must take care to be aware of their own myths when they embark upon the analysis of the myths and stories in organizations. There is a risk that stories will be selectively used to amplify or reinforce the preconceived ideas or assumptions of the researcher when studying narrative in organizations (Gabriel, 2001).

THE AUTHOR'S BIAS

I was a "participant observer," a "fellow traveler," (Patton, 2002) in the research that was the foundation of this book. I chose to study an organization and a group of people that I—as a former Navy Commander—am familiar with. I hope that my familiarity with the Navy yields a richer understanding of the culture than that of a researcher who has not experienced the Navy first hand, but I realize that my background was also the major limitation of this study.

I collected a large amount of raw data in the form of interviews and observations and processed the data for the most part by myself in the development of the findings and conclusions presented in this book. Therefore, the risk of presenting a mono-perspective was unavoidable. However, I took great care to be cognizant of my biases, using protocols to keep my interpretation focused and biases in check. And I engaged in an on-going discourse with most of the participants to check my perceptions. It is hoped that by addressing my background and limitations, the reader will be able to put the findings into perspective. I have also included a great deal of raw data, allowing the storytellers and stories to speak for themselves so that the readers may engage the discourse and draw their own conclusions.

AUTHOR'S BACKGROUND

I retired from the Navy in 1995 as a Commander after 20 years of active service. I thoroughly enjoyed my Navy career, the people I worked with, and the vagabond life style, but I chose to retire because as a wife and mother in a dual career Navy family I felt that my young family needed me, and I saw more family separations in the future if I did not retire. The decision to leave the Navy was the most difficult decision of my life.

The Navy has changed a great deal since I retired. When I joined the Navy women were not yet serving in sea duty assignments. I was never permanently assigned to sea duty, and never spent more than a day at sea on a warship prior to collecting the data for this study. It was exciting for me to spend a week at sea on an aircraft carrier. No doubt my positive experience biased my perceptions. Someone permanently assigned to sea duty, or someone involuntarily assigned to sea duty, would likely paint a different picture of a sea duty environment.

I was stationed in a wide variety of operational, line, and staff positions around the world, throughout my Navy career. Early in my career—at the end of the Vietnam War—I was trained in small group psychotherapy for my assignment as the director of a 40-bed substance abuse treatment facility in Subic Bay, Philippines. This training, and my work with Sailors and Marines who had serious substance abuse problems, increased my interest in oral narratives, enhanced my ability to draw out the narratives of participants, and helped me to be cognizant of group processes as well as narrative content.

My post graduate education in Organization and Leadership, Organizational Development, Strategic Planning, Human Resources Management and Cross Cultural Communication, heightened my awareness of the impact that culture has on narratives and peaked my interest in how to effectively plan for the future and facilitate organizational change. The education and training I received in the Navy, my practical leadership experience, my experience teaching and consulting, and my work as a journalist, gave me a unique perspective that I was able to bring to interpretation of the data I collected for this book.

APPENDIX F

SUGGESTIONS FOR FUTURE RESEARCH AND STUDY

The study of narrative in organizations is relatively new compared to many other areas of social science research. The studies that have been done have only scratched the surface. Those interested in pursuing research in this area have a wide array of questions to explore, such as: Why and how do stories continue to inspire and influence behavior years after they were told? Can stories and narrative inspire altruism, if so, how? Can stories and narratives truly serve to operationalize intuition, if so, how? How can stories be studied to more effectively assess the impact upon listeners? What is the correlation between storytelling skill and leadership success? How can the impact of storytelling be more effectively measured? How can organizations use storytelling to greater effect to problem solve and plan? How can technology be used to support and enhance the development of narrative? Why are some people more effective storytellers than others, and can people learn to become better storytellers? Once stories are part of a deeply ingrained dominant narrative can they be changed, and if so how? How do people decide which stories to tell and when to tell them? What part does luck and serendipity play in determining which stories are told, or are accepted as truth? How can people become more aware of the dominant narratives that are influencing their behavior? How can humans identify, and craft, stories to better communicate, teach and learn? How can attending to story enhance double loop learning? How can story and narrative be used to help those who live and work in chaotic environments—such as war zones—to cope and successfully navigate into the future?

Finally, it would be interesting to explore further how Navy Chief Petty Officers and other effective leaders use story. The narrative put forth in this book only scratched the surface. How might leaders use their experiences, communicated through story and narrative, to enhance the adaptability and understanding of the individuals they are charged with guiding and training? How can leaders use story to bridge the gap between technical knowledge or abstract concepts and real world applications to enhance the adaptability and survivability of their organizations? How can leaders use narrative and story to facilitate organizational change?

BIBLIOGRAPHY

(CAIB), C.A.I.B. (2003). *Columbia Accident Investigation Report*, Washington, D.C.

Barabasi, A.L. (2002). *Linked*, ISBN 0738206679.

Baskin, K. (1998). *Corporate DNA: Learning from Life*, ISBN 0750698446.

Baskin, K. (2008). "Storied spaces: The human equivalent of complex adaptive systems," *Emergence: Complexity & Organization,* ISSN 1521-3250, 10(2): 1-12.

Bettelheim, B. (1970). *The Uses of Enchantment: The Meaning and Importance of Fairy Tales*, ISBN 9780679723936.

Bloch, D.P. (2004). "Spirituality, complexity and career counseling," *Professional School Counseling,* ISSN 1096-2409, 7(5), 343-348.

Bloch, D.P. (2005). "Complexity, chaos and nonlinear dynamics: a new perspective on career development theory," *Career Development Quarterly*, ISSN 0889-4019, 53: 194-207.

Bloch, D.P., Henderson, L.S. and Stackman, R.W. (2007). "Emergence of a social inquiry group: A story of fractals and networks," *World Futures,* ISSN 1556-1844, 63: 194-208.

Bloch, D.P. and Richmond, L.J. (1998). *Soul Work: Finding the Work You Love, Loving the Work You Have,* ISBN 0891061193.

Boje, D.M. (1991). "The storytelling organization: A study of story performance," *Administrative Science Quarterly,* ISSN 0001-8392, *36*(1): 106-127.

Boje, D.M. (1995). "Stories of the storytelling organization: A postmodern analysis of Disney as 'Tamara-land'," *Academy of Management Journal,* ISSN 0001-4273, 38(4): 997-1035.

Boje, D.M. (1998). "Nike, Greek goddess of victory or cruelty? Women's stories of Asian factory life," *Journal of Organizational Change,* ISSN 0953-4814, 11(6): 461-480.

Boje, D.M. (2001). *Narrative Methods for Organizational and Communications Research*, ISBN 0761965866.

Boje, D.M., Fedor, D.B. and Rowland, K.M. (1982). "Myth making: A qualitative step in OD interventions," *Journal of Applied Behavioral Science,* ISSN 0897-3016, 18(1): 17-28.

Boyce, M.E. (1995). "Collective centering and collective sensemaking in the stories of one storytelling organization," *Organization Studies,* ISSN 0170-8406, 16(1): 107-137.

Boyce, M. E. (1996). "Organizational story and storytelling: A critical review," *Journal of Organizational Change,* ISSN 0953-4814, 9(5): 5-26.

Campbell, J. (1973). *The Hero with a Thousand Faces*, ISBN 9781559273305.

Campbell, J. (1988). *The Power of Myth*, ISBN 9780385418867.

Chia, R. (1998). "From complexity science to complex thinking: Organization as simple location," *Organization,* ISSN 1350-5084, *5*(3): 341-369.

Cox, H. (2004). *When Jesus Came to Harvard: Making Moral Choices Today*, ISBN 9780618710546.

Denning, S. (2004). "Telling tales," *Harvard Business Review*, ISSN 00178012, May 2004.

Denning, S. (2005). *A Leaders' Guide to Storytelling: Master the Art and Discipline of Business Narrative*, ISBN 978078797675.

Gabriel, Y. (2000). *Storytelling in Organizations,* ISBN 0198290958.

Gabriel, Y. (2004). *Myths, Stories and Organizations*, ISBN 9780199264483.

Gamache, G., Rosenheck, R. and Tessler, R. (2003). "Overrepresentation of women veterans among homeless women," *American Journal of Public Health,* ISSN 0090-0036, 93(7): 1132-1136.

Gamache, G. and Tessler, R. (2001). "The proportion of veterans among homeless men: a decade later," *Social Psychiatry and Psychiatric Epidemiology,* ISSN 0933-7954, 36: 481-485.

Gell-Mann, M. (1994). *The Quark and the Jaguar: Adventures in the Simple and the Complex*, ISBN 0805072535.

George, J.M. and Jones, G.R. (1996). *Understanding and Managing Organizational Behavior,* ISBN 9780132394574.

Gleick, J. (1987). *Chaos: Making a New Science*, ISBN 0140092501.

Goerner, S.J. (1999). *After the Clockwork Universe*, ISBN 0863152902.

Goleman, D. (1995). *Emotional Intelligence: Why It Can Matter More Than IQ,* ISBN 0553375067.

Hansen, C.D. and William M. (1993). "Storytelling: An instrument for understanding the dynamics of corporate relationships," *Human Relations,* ISSN 0018-7267, 46(12): 1391-2019.

Henshall, S. (2005). *Truth and Story*, from www.henshall.com

Hock, D. (1999). *Birth of the Chaordic Age*, ISBN 1576750744.

Jung, C.G. (1964). *Man and His Symbols*, ISBN 0440351839.

Kauffman, S. (1995). *At Home in the Universe*, ISBN 0195095995.

Lewin, R. (1992). *Complexity: Life at the Edge of Chaos*, ISBN 1587990431.

Lewin, R. and Regine, B. (2001). *Weaving Complexity and Business*, ISBN 1587990431.

Maturana, H. and Varela, F. (1987). *The Tree of Knowledge*, ISBN 0877736421.

Mead, G. H. (1934). *Mind, Self and Society*, ISBN 978022651667.

Patton, M. (2002). *Qualitative Research and Evaluation Methods*, ISBN 0761919716.

Peters, T. and Waterman, R. (1982). *In Search of Excellence*, ISBN 0446378445.

Phillips, D.T. (1992). *Lincoln on Leadership*, ISBN 0446394599.

Ricoeur, P. (1991). *From Text to Action*, ISBN 0810109921.

Ricoeur, P. (1992). *Oneself as Another*, ISBN 0226713288.

Rochlin, G.I., La Porte, T.R. and Roberts, K.H. (1987). "The self-designing high-reliability organization: Aircraft carrier flight operations at sea," *Naval War College Review*, ISSN 0028-1484 (Autumn 1987).

Schein, E.H. (1983). "The role of the founder in creating organization culture," *Organization Dynamics*, ISSN 0957-8234, (Summer): 13-28.

Schein, E.H. (1985). *Organizational Culture and Leadership*, ISBN 0787903620.

Schein, E.H. (1990). "Organizational culture," *American Psychologist*, ISSN 0003-066X, 45:109-119.

Schein, E.H. (2006). "From brainwashing to organizational therapy: A conceptual and empirical journey in search of 'systemic' health and a general model of change dynamics. A drama in five acts," *Organizational Studies*, ISSN 0170-8406, 27: 287-299.

Schein, E.H. and Barker, C.H. (1961). *Coercive Persuasion*, ISBN 0393006131.

Schein, E.H. and Bennis, W.G. (1965). *Personal and Organizational Change through Group Methods: The Laboratory Approach*, ISBN 0471758507.

Stacey, R.D. (1996). *Complexity and Creativity in Organizations*, ISBN 1881052893.

Stacey, R.D. (2000). *Strategic Management and Organizational Dynamics: The Challenge of Complexity*, ISBN 0273708112.

Stackman, R.W., Henderson, L.S. and Bloch, D.P. (2006). "Emergence and community: The story of three complex adaptive entities," *Emergence: Complexity & Organization*, ISSN 1532-7000, 8(3): 78-91.

Waldrop, M.M. (1992). *Complexity: The Emerging Science at the Edge of Order and Chaos*, ISBN 0671767895.

Weick, K.E. (1993). "The collapse of sensemaking in organizations: The Mann Gulch disaster," ISSN 0001-8392, *Administrative Science Quarterly*, 38(4): 628-652.

Weick, K.E. (1995). *Sensemaking in Organizations*, ISBN 0803971761.

Wheatley, M.J. (2006). *Leadership and the New Science*, ISBN 9781576753446.

CPSIA information can be obtained
at www.ICGtesting.com
Printed in the USA
FSHW04n1521090318
45341FS

9 780984 216574